Preservation of Cells

Preservation of Cells

A Practical Manual

Allison Hubel

WILEY Blackwell

Registered Offices
John Wiley & Sons, Inc., 111 River Street, Hoboken, NJ 07030, USA

Editorial Office
111 River Street, Hoboken, NJ 07030, USA

For details of our global editorial offices, customer services, and more information about Wiley products visit us at www.wiley.com.

Wiley also publishes its books in a variety of electronic formats and by print-on-demand. Some content that appears in standard print versions of this book may not be available in other formats.

Library of Congress Cataloguing-in-Publication Data

Names: Hubel, Allison, author.
Title: Preservation of cells : a practical manual / by Allison Hubel.
Description: Hoboken, NJ : John Wiley & Sons, 2018. | Includes index.
Identifiers: LCCN 2017042796 | ISBN 9781118989845 (cloth) | ISBN 9781118989876 (epub) | ISBN 9781118989852 (epdf)
Subjects: LCSH: Cells–Cryopreservation–Handbooks, manuals, etc. | Cells–Preservation–Handbooks, manuals, etc.
Classification: LCC QH324.9.C7 H83 2018 | DDC 571.6/34–dc23
LC record available at https://lccn.loc.gov/2017042796

Cover Design and Image: Created by Alex Brown

Set in 10/12pt Warnock by SPi Global, Pondicherry, India
Printed and bound in Malaysia by Vivar Printing Sdn Bhd

10 9 8 7 6 5 4 3 2 1

Contents

Protocols

Preface

Preservation of cells is performed thousands of times every day by technicians across the world. For the vast majority of those people, the process is shrouded in mystery. The reasons for specific steps in the protocol are not clear. If there are problems with the protocol, the manner by which the problems occur is also unclear. There are numerous books that contain cryopreservation protocols for specific cell types or describe scientific understanding of the field or current research in the field. I could not find a book, which helped many, that uses preservation to develop new protocols or improve existing protocols. The objective of this book is to describe step by step the development of a preservation protocol and the scientific principles behind these steps. At the end of every chapter (with the exception of Chapter 8), specific links are made between scientific principles and the manner by which those principles are put into action.

Cells are being used for an increasing number of downstream uses. The applications of cells include the production of therapeutic proteins, viral vaccines, and antibodies. Cells are being used as biomarkers for health and disease and even for the treatment of disease. These applications are described in Chapter 1 and the role of preservation in the clinical and commercial applications of cells is also described. Different modes of preservation are also described. Different applications of cells may involve hypothermic storage of cells, cryopreservation, or vitrification.

Cells undergo a variety of processes prior to cryopreservation. These processes can include digestion from a tissue, selection of subpopulations, genetic modification, culture, and so forth. Chapter 2 describes these processes in more detail and the resulting nutrient deprivation, shear, or other sublethal stresses that can influence post-thaw recovery of cells. Strategies to minimize stress or cell losses from pre-freezing processes are described. Newly developed gene-editing technologies are described. The ability to edit cells may lead to both new challenges and opportunities for cell preservation. It is likely that insertion or deletion of specific genes may influence the ability of a cell to survive the stresses of freezing and thawing. Gene editing may also enable us to understand the role of specific genes in enhancing survival of certain cells.

Characterization of the cells being cryopreserved is also critical. Chapter 2 also describes standard testing of cells prior to cryopreservation, including identification of cells, testing for adventitious agents, and other types of testing such as genetic stability. Misidentification of cells is a serious concern in the area of life science research, and there is increasing emphasis on proper identification of both primary cells and cell lines being used.

Cryopreservation uses specialized solutions designed to help the cells survive the stresses of freezing and thawing. These solutions are not physiological. Chapter 3 describes formulation of a solution and development of methods to introduce the solution. A listing of molecules that have been used to stabilize cells during freezing is given in the chapter. The development of new solutions is an ongoing area of research, but these solutions will still, more than likely, need to be introduced and removed prior to downstream use.

The influence of cooling rate on the post-thaw survival of cells has been known for almost 50 years. Chapter 4 describes the cooling process and the manner by which cells are typically frozen (i.e., controlled-rate freezing, passive freezing, or vitrification). Designing a cooling protocol and methods of verifying the protocol are also described. The importance of temperature and its variation in time during freezing also suggests that the method of measuring temperature independently during freezing is valuable, in particular during the development of methods.

Cells that have been cryopreserved may be stored for weeks, months, or even decades. Chapter 5 describes the scientific basis for storage of cells in liquid nitrogen, fundamentals of repository design, safe operation of a repository, and shipping of samples from the site of storage to the site of use. The factors that influence stability of samples in storage are also discussed. Transient warming events (TWEs) are being documented for a wide variety of biospecimens in storage and our understanding of the influence of TWEs on sample quality continues to grow. It is likely that new technologies can be used to eliminate this issue and improve stability of samples in storage.

The purpose of preservation is to maintain the critical biological properties for downstream use of the cells; downstream use of the cells requires thawing of the sample. The thawing process and the manner by which you can characterize your average thawing rate and improve the thawing process are described in Chapter 6. Newly developed controlled-thawing technology will provide the opportunity to improve the consistency of thawing. In addition, new types of thawing protocols may be developed in the future, which will improve overall outcome.

It is common for cells to be washed post-thaw and prior to downstream applications. Methods and technologies for washing cells post-thaw are also described in Chapter 7. For vitrification solutions or cells that are sensitive to osmotic stress, strategies for improved methods of washing are described.

Effective methods of preserving cells cannot take place without effective methods of characterizing post-thaw recovery. Post-thaw assessment of cells is a very common area for errors and poor practices. The need for post-thaw function of cells, in particular for cells used therapeutically, implies that post-thaw function is critical and methods of assessment must be meaningful. Different methods of post-thaw assessment are described. Specific recommendations are given to reduce bias and errors.

The traditional method of optimizing a preservation protocol typically involves empirical testing (i.e., varying composition and cooling rate and measuring post-thaw viability). Chapter 8 describes the use of a differential evolution algorithm to reduce the experimentation required to optimize composition, cooling rate, and other processing parameters for cells. As there is pressure to develop fit-for-purpose protocols, new methods to streamline the cost and time required for optimization are critical and this approach has the potential to be transformative.

The growth in clinical and commercial applications of preservation brings with it the need for consistency and reproducibility. Chapter 1 describes common errors in preservation practices that lead to poor reproducibility (both poor outcome and high variability). Subsequent chapters describe common pitfalls that can negatively affect reproducibility of the preservation process and strategies to avoid those pitfalls and improve reproducibility.

For all the reasons listed previously, conventional methods of cryopreservation may no longer be appropriate for a given cell type or a given application. Drift in preservation protocols is also common. As director of the Biopreservation Core Resource, I get phone calls and emails from organizations that start having problems with existing protocols. The overall goal of this book is to help both groups. The process of developing a new protocol or understanding problems with an existing protocol can be approached logically and systematically based on scientific principles. It is my hope that this book will enable more organization to achieve improved post-thaw recoveries and consistency.

Acknowledgments

This book grew out of a short course, "Preservation of Cellular Therapies," offered at the University of Minnesota for well over a decade. Dave McKenna, Fran Rabe, and Diane Kadidlo from the Molecular Cellular Therapy Program at the University of Minnesota helped me understand the complexities of preserving cells in a clinical context, regulatory issues, and the importance of quality systems in preservation.

Ian Pope from Brooks Life Sciences helped structure the chapter on storage and his critical reading of the manuscript helped me understand the importance of directly linking the scientific principles to actual practice. Amy Skubitz brought her decades of experience in biobanking and her critical eye to the manuscript as well. Her insights made the book far better and for that I am grateful. Alex Brown contributed the wonderful illustrations and his artistic eye to the project.

I would also like to thank all of the protocol contributors: Leah A. Marquez-Curtis, A. Billal Sultani, Locksley E. McGann, and Janet A. W. Elliott from the University of Alberta; Rohit Gupta and Holden Maeker from Stanford University; Melany Lopez and Ali Eroglu from the Medical College of Georgia; Andreas Sputtek from Medical Laboratory Bremen; Jeffrey Boldt from Community Health Network; and Jerome Ritz, Sara Nikiforow; and Mary Ann Kelley from Dana Farber Cancer Institute, Boston, MA, USA. All of these protocols are excellent examples of putting the scientific principles described in the book into practice.

Finally, thank you to John Martin Hansen, my husband, for his patience and support through this process.

Nomenclature

ΔT	Undercooling of the cells
AABB	American Association of Blood Banks
AATB	American Association of Tissue Banks
ALP	Alkaline phosphatase
B	Cooling rate
CR	Crossover rate
DMSO	Dimethylsulfoxide
DOT	Department of Transportation
DSC	Differential Scanning Calorimetry
ES cells	Embryonic stem cells
F	Weighting
FACT	Foundation for the Accreditation of Cellular Therapy
FDA	Food and Drug Administration
GMP	Good Manufacturing Practices
HIV	Human immunodeficiency virus
HSC	Hematopoietic stem cells
IATA	International Air Transport Association
ICAO	International Civil Aviation Organization
iPS cells	Induced pluripotent stem cells
ISBER	International Society for Biological and Environmental Repositories
k	Interaction parameter
LN_2	Liquid nitrogen
MSC	Mesenchymal stromal cells
NP	Generation size
PBPC	Peripheral blood progenitor cells
R	Rate constant
RBCs	Red blood cells
RNA A	Ribonuclease A
T_{ext}	Temperature at which ice forms in the extracellular solution
T_{final}	Final temperature

t_{final}	Final time
T_g	Glass transition temperature
$T_{initial}$	Initial temperature
$t_{initial}$	Initial time
T_m	Melting temperature
T_{nuc}	Nucleation temperature
UCB	Umbilical cord blood
x	Weight fraction

1

Introduction

Mammalian Cells: Modern Workhorses

Mammalian cells have become modern workhorses capable of a variety of applications:

- Production of therapeutic proteins, viral vaccines, and antibodies
- Therapeutic agents (cell therapy or regenerative medicine applications)
- Biomarkers for health or disease
- *In vitro* models (i.e., replacement for animal testing)

These applications represent significant economic sectors and have a major impact on human health.

Products from Cells

The production of human tissue plasminogen activator (tPA) in the mid-1980s became the first therapeutic protein derived from mammalian cells to be made available commercially (see Wurm (2004) for review). Erythropoietin, human growth hormone, interferon, human insulin, and a variety of other proteins are produced from mammalian cells and are used therapeutically. Since the production of tPA, roughly 100 recombinant protein therapeutics have been approved by the FDA (Lai, Yang, and Ng 2013).

In addition to therapeutic proteins, vaccines are commonly produced from mammalian cells. For example, polio, hepatitis B, measles, and mumps vaccines are all produced via mammalian cell culture. New vaccines currently under development (human immunodeficiency virus (HIV), Ebola, new influenza strains) are also based on mammalian cell cultures.

Antibodies are used for a wide range of application (both *in vitro* and *in vivo*) (Waldmann 1991). The diagnosis of disease using antibodies is an extremely common application. Enzyme-linked immunoabsorbent assays, flow cytometry, immunohistochemistry, and radioimmunoassays all use monoclonal antibodies produced by mammalian cells. Clinical applications of antibodies have historically included treatment of viral infections. Immunotherapy for the treatment of cancer using antibodies has grown rapidly (Weiner, Surana, and Wang 2010). Antibodies are now being used to selectively target tumors. The ability to accomplish targeting of tumors in humans resulted directly from advances in antibody engineering that enabled production of chimeric, humanized, or fully human monoclonal antibodies. Antibodies have also been conjugated to drugs or radioactive isotopes and used as target therapies. Currently, more than 10 different antibodies are approved for the treatment of cancer. All of these antibodies are produced using mammalian cells.

Cells as Therapeutic Agents

Cell therapy began in the 1970s with bone marrow transplantation for the treatment of blood and immune disorders. The uses of hematopoietic stem cells (HSCs) have grown since then, and clinical studies are expanding the use of HSCs to include a wider range of diseases and indications. Over 430 clinical trials using HSCs are underway, targeting the immune system, cardiovascular diseases, neurological disorders, vascular disease, lung disease, and HIV, to name a few (Li, Atkins, and Bubela 2014). The discovery that HSCs can be found in the peripheral blood (if a patient has been given a drug to mobilize HSCs to circulate in the peripheral blood), and umbilical cord blood (UCB) has enabled growth in the use of this cell type therapeutically because these cells can be harvested using nonsurgical methods.

Stromal cells present in the bone marrow microenvironment have also been studied for therapeutic uses. Mesenchymal stromal cells (MSCs) provide important support for hematopoiesis in the bone marrow microenvironment. MSCs can also be isolated from adipose tissue and UCB. Initial studies using MSCs focused on regenerative medicine applications and the use of these cells to form bone or cartilage. Subsequent studies demonstrated that the principal actions of MSCs are immunomodulatory and trophic (Caplan and Correa 2011). The diverse capabilities of this cell type and the ability to access the cells easily (from bone marrow aspirate, UCB, or small biopsy of adipose tissue) have facilitated clinical use of these cells. Clinical trials use MSCs to treat orthopedic disorders, cardiovascular disease, autoimmune disease, neurological disorders, and more (Sharma et al. 2014). MSCs are immune privileged, and as a result cells from allogeneic donors can be given therapeutically.

Biomarkers for Health or Disease

Most people have had a vial of blood drawn at the doctor's office. Blood counts are performed and can indicate the presence of anemia, infection, or other medical conditions. The cells in whole blood are typically not stored for an extended period of time but counted shortly after collection. Other cell-based assays include quantification of circulating tumor cells as a marker of tumor burden in cancer (Plaks, Koopman, and Werb 2013). Flow cytometry of lymphocyte subsets is also used to monitor immune status for AIDs patients and others with immune disorders (Shapiro 2005).

Mass cytometry is a recently developed experimental technique in which heavy metals are tagged to antibodies and those labels are attached to cells. The cells are then analyzed using a time-of-flight mass spectrometer. This approach avoids the limitations intrinsic to conventional flow cytometry. This capability enables labeling of heterogeneity in a cell population as well as single cell analysis of multiple markers (Spitzer and Nolan 2016). Other single cell "omic" (genomic, proteomic, and metabolomic) technologies are in development and could represent powerful new diagnostics. It is likely that cells will continue to grow in importance for diagnostics.

In Vitro Models

For many years, isolated hepatocytes have been used for screening of drugs. The development of induced pluripotent stem cells (iPS cells) (Yu et al. 2007) and the ability of these cells to differentiate into a variety of cell types have enabled the testing of drugs in a wider variety of cell types. For example, the cardiotoxicity of a drug can be evaluated using cardiomyocytes differentiated from iPS cells (Avior, Sagi, and Benvenisty 2016).

Three-dimensional cultures of multiple cell types in a microfluidic environment that permits continuous perfusion of the cells can be used to model the physiological function of an organ or tissue. Also known as "Organ-on-a-chip," these cultures are also being used to screen drugs and understand the effect of specific drugs on organ systems (Bhatia and Ingber 2014).

Organ-on-a-chip and iPS cells are also used for modeling of disease. Cells from donors with a given disorder can be transformed into iPS cells and then differentiated into disease-specific cells that can be used for understanding disease development as well as drug/treatment screening (Avior, Sagi, and Benvenisty 2016). Patient-derived iPS cells have been developed for a wide range of diseases, including neurological disorders and cardiovascular disease.

Clearly, mammalian cells are critical for biomedical research, diagnosis of disease and its treatment. These cells must be functional and available at the site and at the time of downstream use.

Bridging the Gap

It is common for cells to be collected or cultured in one location and used at a later time and in another location (Figure 1.1). The critical biological properties of the cells must be preserved in order for the cells to be useful for the downstream applications.

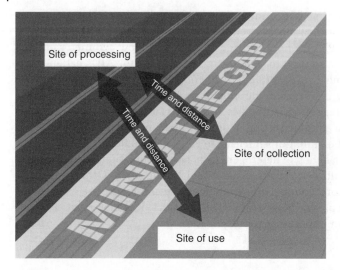

Figure 1.1 Minding the gap: preservation is used to maintain the critical biological properties of the cells when they are needed at a later time or in a different location.

Cells that are to be used therapeutically must be properly stored to meet the safety and quality control testing prior to release (and use) of the cells. Preserving cells permits coordination of the therapy with patient-care regimes (i.e., the cells are ready when the patient is ready). Cells used therapeutically are produced in specialized facilities. The ability to preserve cells will help manage staffing requirements for cell-processing facilities (i.e., the cells can be processed independent of patient availability) and control the inventory of the therapy.

UCB banking is a good example of the need for preservation. Babies are born at unpredictable times and at a variety of locations. The UCB must be collected immediately after birth. The UCB is typically shipped directly to a cord blood–processing facility where the sample is depleted of red blood cells, cryopreserved, and stored. The unit is stored until it is needed (typically years later), and it is common for the unit to be used in a third location. The UCB unit is useful only if the critical biological properties have been preserved. The genetic diversity of births implies that the preservation of UCB, in particular, can improve the genetic diversity of cells available for therapeutic applications.

One option for preserving a cell may be keeping the cells in culture until they are ready for use. Certain cell types do not retain their critical biological characteristics if they are cultured outside the body for extended periods of time. Other cell types that have defined genetics (e.g., mammalian cells used for the production of recombinant proteins) may experience genetic drift with long-term culture. Finally, long-term culture can be very expensive. For many of the cells described above, the downstream use of the cells may be months or weeks after the cells are collected. As a result, cryopreservation is a useful tool for preserving the critical biological properties of the cell for extended periods of time.

The Preservation Toolkit

There are a variety of methods that can be used to preserve the cell depending upon its downstream application (Figure 1.2). Multiple modes of preservation may also be used in a given protocol. Using the example given above, UCB is collected in the delivery room, but it is processed in specialized facilities for cord blood banking. The cells are shipped using short-term liquid storage to a centralized facility where they are cryopreserved. Therefore, this particular application uses liquid storage followed by cryopreservation. Each mode has its advantages and limitations.

Hypothermic Storage

Hypothermic storage is commonly used for short-term (hours to days) storage of cells. Cells are taken as collected or resuspended in a storage solution and typically refrigerated or placed on ice (hence the term hypothermic storage). Reducing the temperature of cells reduces their metabolic activity, enabling the

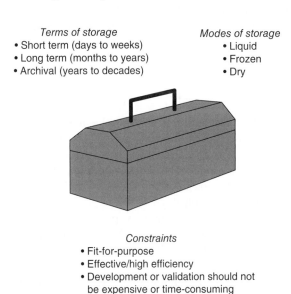

Terms of storage
• Short term (days to weeks)
• Long term (months to years)
• Archival (years to decades)

Modes of storage
• Liquid
• Frozen
• Dry

Constraints
• Fit-for-purpose
• Effective/high efficiency
• Development or validation should not
  be expensive or time-consuming

Figure 1.2 Preservation toolkit for cells. Preservation methods should be fit-for-purpose. The mode and duration of storage should be appropriate for the given application.

cells to be shipped or transported. It is noteworthy that when refrigerated, the cells are still consuming oxygen and other nutrients. Storage conditions (e.g., temperature, time, and duration) must be chosen such that the cells are functional at the completion of the liquid storage.

Red blood cells (RBCs) are the most common cell type that is stored using this method. Red blood cells are separated from whole blood and resuspended into a specially designed short-term storage solution (e.g., AS-3) and refrigerated. RBCs can be stored up to 42 days in this solution. Since they do not replicate, the ability of RBCs to be stored for this period of time in liquid storage reflects the unique biology of this cell type. Most nucleated cells cannot be stored for this period of time, even when refrigerated.

At reduced temperatures, ion pumps in the cell membrane do not function properly and there is a change in ionic concentration inside the cell. Low temperatures also influence mitochondrial activity (e.g., reduced ATP production and diminished free radical scavenging). It has been hypothesized that damage during hypothermic storage results from reactive oxidative species (Rauen and de Groot 2002). Even with specially designed solutions, liquid storage of nucleated cells before significant losses occur is typically limited to short periods of time (less than 72 h).

Liquid storage can be used in combination with a cryopreservation protocol. For example, UCB is typically collected in the labor and delivery room and shipped to a cord blood bank. It is common for the cells to be chilled (on ice) but not frozen and then processed at the UCB bank within a short (approximately 24 h) period of time after collection. Studies have shown that improper liquid storage conditions can result in poor post-thaw recovery of cord blood (Hubel et al. 2004).

Cryopreservation

The use of freezing to stabilize biological cells results from the need to control or stop degradative processes. Specifically, during freezing, liquid water is removed from the sample in the form of ice. Liquid water is a critical component in a variety of metabolic functions of the cell. Freezing of the water in the sample reduces the mobility of water molecules and therefore their ability to participate in reactions that could potentially degrade the cell.

All cells contain degradative enzymes (e.g., DNAses, proteases, etc.). The activity of these enzymes is a function of temperature. As the temperature is reduced, the activity of the enzymes decreases and there is a threshold at which the enzyme is no longer active and therefore cannot participate in degradation of the cell. Activity of a limited number of enzymes has been measured at freezing temperatures, and these studies suggest that the threshold temperatures for activity may be approximately −90°C (Hubel, Spindler, and Skubitz 2014).

Cells that have been successfully frozen can be stored for longer periods of time (years to decades), thereby extending the shelf-life of the product. The process requires maintaining a cold chain (a temperature controlled supply chain that keeps the product at a desired low temperature) during freezing, storage, and transport.

Vitrification

As described above, water is removed from the sample in the form of ice during conventional cryopreservation. As a result, significant changes in the chemical and mechanical environment of the cells take place. When water is removed in the form of ice, the remaining unfrozen solution contains high concentration of solutes. The cells are sequestered in gaps between adjacent ice crystals and are therefore subjected to high concentrations and mechanical forces as freezing progresses.

One approach to preserving the cells at low temperatures, known as vitrification, involves avoiding the formation of ice. Typically, high concentrations of cryoprotective agents are used to suppress ice formation during freezing and, in fact, concentrations can be considerably higher than used in conventional cryopreservation. These solutions are not physiological and, typically, the process of adding these solutions to a biological system is called introduction of the solution. Introduction and removal of these high-concentration solutions may be elaborate and require multiple steps for introduction. For example, if the final concentration of the solution is 4 M and the cell cannot tolerate introduction of the solution in a single step, the cell may be introduced to a solution whose concentration is intermediate (say 1.2 M) followed by introduction to the solution at its final concentration (4 M). The process is designed to reduce cell losses resulting from osmotic stress (see Chapter 2 for more details). Vitrification solutions may also contain additives designed to mitigate the toxicity of the cryoprotective agents.

Cooling of a sample to be vitrified typically involves plunging the sample in liquid nitrogen (LN_2) in order to achieve the fastest cooling rate possible. Samples that are vitrified are typically stored at temperatures near the glass transition temperature of the sample, T_g. T_g is the temperature (or range of temperatures) at which a solution forms an amorphous phase. Few studies have examined the stability of vitrified samples. Recrystallization (i.e., the nucleation and growth of ice) can be observed during storage, which will affect the stability of the product in the long term. Thawing of the sample requires warming rates that are approximately two orders of magnitude faster than the cooling rate of the sample. The complexity of the process (i.e., introduction and removal of high concentration solutions, rapid freezing, and potential for damage during storing and warming) means that this technique is not commonly used. Gametes and embryos are the most common samples that are vitrified clinically. Protocols for these cell types involve small volumes, and the techniques for physical manipulation of the samples are widely known in the field, which facilitates use of vitrification.

Dry State Storage

Currently, DNA is the most common sample stored in dry form (Ivanova and Kuzmina 2013). An aqueous sample containing DNA is placed in a matrix typically containing sugars and other additives. Water is removed from the sample by drying until the sample becomes amorphous. On a basic level, dry state storage is similar to vitrification in that an amorphous phase is formed. In conventional vitrification, cold temperatures are used to form an amorphous phase. For dry state storage, the combination of water content and solute forms an amorphous phase. In contrast to cryopreserved samples, dry state storage does not require a low temperature environment. It does, however, require control of humidity. Uptake of water by the sample will result in degradation. At present, cells cannot be stored in the dry state.

Fit-for-Purpose

As described previously, cells may be collected in a given time and location only to be used at another time and location. The site of collection may be a hospital room, a battlefield, or even a coral reef. The downstream use of the sample may be hours, days, or decades later and, as described previously, may involve a range of uses. The preservation protocol used must also fit into the workflow and will vary with the following:

- The needed duration of storage (hours, days, weeks, years)
- Downstream use (i.e., quality of sample required)
- Training or skill level of operator
- Resources required and resources available
- Cell type

As a result, the preservation protocol as a whole must be fit-for-purpose. The preservation toolkit described previously can be used to develop fit-for-purpose preservation protocols.

For example, mononuclear cells obtained from peripheral blood (PBMNCs) have a variety of downstream uses, and each use may require a different preservation processing technique. PBMNCs may be used for the isolation of DNA. The cells themselves do not need to be viable and, in most situations, the cells are processed to form a pellet and the pellet is stored in a manner to maintain the integrity of the DNA. Another application of cryopreserved PBMNCs is to recover cells and immortalize them in order to create a donor-specific cell line.

Enough cells must survive the freezing process to result in successful immortalization. Finally, PBMNCs may be used therapeutically. Lymphocytes (mixed or specific subsets) can be given to patients in order to achieve a therapeutic outcome (e.g., graft versus leukemia effect, modulation of immune response, tumor fighting capacity). In this situation, preservation has to be highly efficient (e.g., minimal cell losses), and a post-thaw function of the cells is required.

• *Development of a preservation protocol/process requires an understanding at the outset of the downstream purpose of the cell.*

Every step of the process should reflect back upon the desired downstream use of the cell. The preservation toolkit can be used to modify the protocol for the desired downstream use. Each element of the protocol is based on scientific principles and those principles can be used to rationally design that particular step to achieve the desired downstream use of the cell.

One Size Does Not Fit All

For researchers, clinicians, or biorepository operators, a single universal cryopreservation protocol appropriate for every cell type has been a goal. The use of a 10% dimethylsulfoxide (DMSO) solution and a cooling rate of 1°C/min is to some a "universal preservation protocol." Unfortunately, the use of DMSO is not appropriate for every downstream application, and there are several cell types of tremendous value that are not effectively preserved using that protocol.

The simple reality is that the freezing response of a cell is driven by biophysics and biology. The hematopoietic system is an excellent example of differences in biology influencing freezing response. Hematopoietic progenitor cells (HPCs) are routinely cryopreserved using 10% DMSO and a cooling rate of 1°C/min. The mature blood cells that are derived from HPCs have a range of freezing responses. RBCs are cryopreserved in either 17 or 40% glycerol. Platelets and neutrophils cannot be effectively preserved using any conventional methods. Certain subpopulations of lymphocytes can be effectively preserved using DMSO whereas others are not. Freezing responses can vary amongst different cell types and also from species to species. For example, the freezing behavior of sperm varies from species to species. Cryopreservation protocols must reflect the unique characteristics of the cell type being preserved.

The Process is the Product

Unlike a lot of other bioprocesses, the properties of the end product (the viability, recovery, and function of the cells post-thaw) represents the cumulative effect of all of the processing steps and all of the reagents used in the process (Figure 1.3).

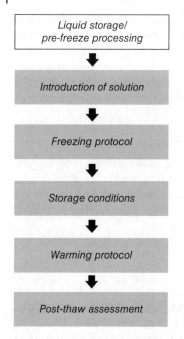

Figure 1.3 The different steps of the preservation process. Each step of the process can be designed based on scientific principles. Each step contributes to the overall quality of the end product.

Cells may be subjected to a variety of processes prior to freezing: culture, selection of subpopulations, genetic modification, and more. These processes may be stressful to the cells and influence their ability to survive the additional stresses of freezing and thawing (Chapter 2).

Cells are frozen in specialized solutions designed to help them survive the stresses of freezing and thawing (Chapter 3). Cells cannot be sterilized or purified after processing, so anything added to the sample remains in the sample. This consequence implies that reagents used in the preservation process must be defined and must be of high quality (Figure 1.4).

Cryopreservation solutions are not physiological, and introducing the solution or exposing cells to the solution can result in cell loss/death independent of the freezing process.

The freezing process requires taking a biological system normally at physiological temperatures (approximately 37°C) and traversing a wide temperature range to cryogenic temperatures (−196°C). The manner by which this is done (most often described by the cooling rate) plays a strong factor in the survival of the cells (Leibo and Mazur 1971) (Chapter 4). Controlled-rate freezers or passive freezing devices are used to control the rate at which cells traverse that temperature difference. Protocols for controlled cooling of samples can be complex and involve multiple processes. Each process (described most often as a "segment") can be rationally

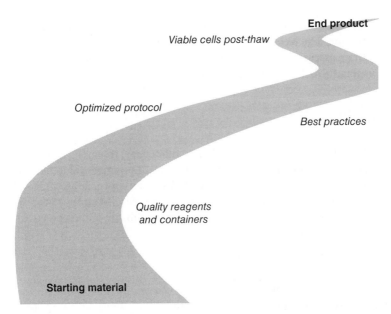

Figure 1.4 The reagents, starting material, and processes determine the quality of the product at the end of preservation.

designed and optimized. The temperature at which ice forms in the extracellular solution can influence the post-thaw viability of cells frozen in a given solution at a given cooling rate. Different strategies are described for controlling (or not) that temperature.

After freezing, the samples are placed in a repository for long-term storage. Proper storage temperatures can be selected based on the composition of the cryopreservation solution and the cells being stored (Chapter 5). Accessing the repository to add or remove samples can influence the temperature of a sample and therefore its post-thaw recovery. It is common for the samples to be used at a different location and the samples may be shipped to that location. Maintaining a low temperature during shipping is critical to maintaining the quality of the samples.

When warming the sample, the same range of temperatures traversed during freezing must once again be traversed in a manner that does not damage the sample. In general, the warming rate must be more than an order of magnitude greater than the rate of cooling and 60–80°C/min is preferable (Chapter 6). As cryopreservation solutions are not physiological solutions, the samples may need to be washed or diluted post-thaw prior to the downstream use of the cells.

The purpose of preservation is to retain the critical biological properties of the sample. A protocol must also include proper methods for post-thaw assessment. Specifically, the method of post-thaw assessment must reflect the downstream uses of the cells and often involves more than just the integrity of the cell membrane. The freezing process brings about changes in the cells that make post-thaw assessment challenging. Care must be used when selecting and interpreting the behavior of the cells post-thaw (Chapter 7).

Reproducibility

Poor reproducibility is a significant problem in life science research (Freedman et al. 2015), resulting in increased delays and cost of therapy development. It has been estimated that the lack of reproducibility costs approximately US$28 billion/year. Specific factors have been identified that resulted in poor reproducibility of cryopreservation protocols (Freedman, Gibson, and Neve 2015). These factors include the following:

- Cells were overstressed during preservation
- Freezing/thawing media was incorrect
- Freezing media components had expired
- Cells were stored or cryopreserved incorrectly
- Cells were thawed incorrectly
- Too few cells were frozen

Preservation protocols are not immune to concerns regarding rigor and reproducibility. Throughout this book, specific practices are described that are intended to improve both the outcome (i.e., cell recovery and function) and the reproducibility of preservation protocols.

Safety

Preservation processing may expose workers to specific hazards. For example, handling of solutions containing DMSO is associated with specific hazards (rapid transport through the skin and permeabilization of skin which can in turn lead to adverse events). Storage of cells on LN_2 can bring with it both exposure to cold temperatures and the potential for asphyxiation. A description of the hazards as well as strategies for dealing with these potential hazards is addressed in the relevant chapters (Chapters 3 and 5).

Dispelling the Myth of the Cold Black Box

Preservation of cells is performed on a daily basis by tens of thousands of technicians. For the majority of these people, the preservation protocol is a cold black box, which is not clearly understood. Achieving the desired outcome from a preservation protocol is only possible when the end user understands that a preservation protocol is built out of individual steps (Figure 1.3) and each of those steps can be constructed using scientific principles. The information in this book will enable users to both construct a new protocol and debug an existing protocol to improve outcome. In order to make the links between the scientific principles and the structure of the protocol, each chapter contains a summary at the end, which describes the scientific principles and practical tips for putting those principles into practice.

References

Avior, Y., I. Sagi, and N. Benvenisty. 2016. "Pluripotent stem cells in disease modelling and drug discovery." *Nat Rev Mol Cell Biol* 17 (3):170–182.

Bhatia, S. N., and D. E. Ingber. 2014. "Microfluidic organs-on-chips." *Nat Biotechnol* 32 (8):760–772.

Caplan, A. I., and D. Correa. 2011. "The MSC: an injury drugstore." *Cell Stem Cell* 9 (1):11–15.

Freedman, L. P., M. C. Gibson, S. P. Ethier, H. R. Soule, R. M. Neve, and Y. A. Reid. 2015. "Reproducibility: changing the policies and culture of cell line authentication." *Nat Methods* 12 (6):493–497.

Freedman, L. P., M. C. Gibson, and R. M. Neve. 2015. "Changing the culture of cell culture: applying best practices and authentication to ensure scientific reproducibility." *Biopharm Int* 28 (10):14–20.

Hubel, A., D. Carlquist, M. Clay, and J. McCullough. 2004. "Liquid storage, shipment, and cryopreservation of cord blood." *Transfusion* 44 (4):518–525.

Hubel, A., R. Spindler, and A. P. Skubitz. 2014. "Storage of human biospecimens: selection of the optimal storage temperature." *Biopreserv Biobank* 12 (3):165–175.

Ivanova, N. V., and M. L. Kuzmina. 2013. "Protocols for dry DNA storage and shipment at room temperature." *Mol Ecol Resour* 13 (5):890–898.

Lai, T., Y. Yang, and S. K. Ng. 2013. "Advances in Mammalian cell line development technologies for recombinant protein production." *Pharmaceuticals (Basel)* 6 (5):579–603.

Leibo, S. P., and P. Mazur. 1971. "The role of cooling rates in low-temperature preservation." *Cryobiology* 8 (5):447–452.

Li, M. D., H. Atkins, and T. Bubela. 2014. "The global landscape of stem cell clinical trials." *Regen Med* 9 (1):27–39.

Plaks, V., C. D. Koopman, and Z. Werb. 2013. "Cancer circulating tumor cells." *Science* 341 (6151):1186–1188.

Rauen, U., and H. de Groot. 2002. "Mammalian cell injury induced by hypothermia—the emerging role for reactive oxygen species." *Biol Chem* 383 (3–4):477–488.

Shapiro, H.M. 2005. Practical Flow Cytometry. Hoboken, NJ: John Wiley & Sons, Inc.

Sharma, R. R., K. Pollock, A. Hubel, and D. McKenna. 2014. "Mesenchymal stem or stromal cells: a review of clinical applications and manufacturing practices." *Transfusion* 54 (5):1418–1437.

Spitzer, M. H., and G. P. Nolan. 2016. "Mass cytometry: single cells, many features." *Cell* 165 (4):780–791.

Waldmann, T. A. 1991. "Monoclonal antibodies in diagnosis and therapy." *Science* 252 (5013):1657–1662.

Weiner, L. M., R. Surana, and S. Wang. 2010. "Monoclonal antibodies: versatile platforms for cancer immunotherapy." *Nat Rev Immunol* 10 (5):317–327.

Wurm, F. M. 2004. "Production of recombinant protein therapeutics in cultivated mammalian cells." *Nat Biotechnol* 22 (11):1393–1398.

Yu, J., M. A. Vodyanik, K. Smuga-Otto, J. Antosiewicz-Bourget, J. L. Frane, S. Tian, J. Nie, G. A. Jonsdottir, V. Ruotti, R. Stewart, Slukvin, II, and J. A. Thomson. 2007. "Induced pluripotent stem cell lines derived from human somatic cells." *Science* 318 (5858):1917–1920.

2

Pre-freeze Processing and Characterization

What happens to cells immediately before they are frozen influence their ability to survive the stresses of freezing and thawing. Techniques for manipulating cells have grown in tandem with their uses. Examples of common methods of processing or manipulating cells include the following:

- Digestion of cells from an intact tissue or organ
- Hypothermic storage
- Selection of subpopulations of cells (depletion or enrichment)
- Activation or stimulation of cells
- Genetic modification (insertion or deletion of genes)
- Culture

Cells that experience these processes can be exposed to physical stresses (e.g., centrifugation, fluid shear, etc.), nutrient deprivation, and other stresses such as infection by a virus or electroporation. These pre-freeze stresses are in general sublethal but can influence the ability of the cells to survive the subsequent stresses of freezing and thawing. This chapter is intended to help the reader understand the influence of processing procedures on the health of cells prior to cryopreservation and strategies to mitigate stress on the cells.

Pre-freeze processing should also involve characterization of the sample being preserved. At minimum, the identity of the cells should be confirmed and the sample should be tested for the presence of adventitious agents (e.g., viruses, mycoplasma, etc.). Cell therapies will require additional pre-freeze characterization and those requirements will be discussed as well.

Pre-freeze Processing

Digestion of Cells from Intact Tissue

A common method of isolating primary cells embedded in tissue or organs involves digesting the extracellular matrix to remove the cells. The tissue may

Preservation of Cells: A Practical Manual, First Edition. Allison Hubel.
© 2018 John Wiley & Sons, Inc. Published 2018 by John Wiley & Sons, Inc.

be mechanically disrupted and/or exposed to enzymes designed to break down the extracellular matrix and dissociate the cells. Two different types of stress result from this process: hypoxia resulting from cutting off the blood supply to the tissue and stress associated with enzymatic digestion of the tissue. The effect of these pre-freeze processes has been studied most extensively in hepatocytes. Hepatocytes immediately postharvest (Hubel, Conroy, and Darr 2000) are extremely hypoxic. Anoikis (programmed cell death resulting from detachment) is common in hepatocytes and a source of cell losses both during culture and cryopreservation (Nyberg et al. 2000, Yagi et al. 2001). A brief incubation period in oxygenated conditions is the most common method of helping cells recover from the stresses of isolation from tissues or organs (Li et al. 1999).

Hypothermic Storage

It is common for cells to be chilled (typically on ice) and shipped or transported. Several changes take place when cells are held at low (but not freezing) temperatures. Reducing the temperature of the cell reduces its metabolic activity. Similarly, ion pumps in the cell membrane do not function properly and ionic concentration inside the cell can change. A common manifestation of this effect results in swelling of the cells. Low temperatures also influence mitochondrial activity (e.g., reduced ATP production and diminished free radical scavenging). It has been hypothesized that damage during hypothermic storage results from reactive oxidative species (Rauen and de Groot 2002). Hypothermic storage of nucleated cells without significant losses is typically limited to short periods of time (less than 72 h).

It is common for cells collected from living donors (human or animal) to use plasma as a short-term storage solution. Plasma contains many of the nutrients needed for short-term storage and in general is a healthy environment for cells. From normal, healthy donors, the use of plasma or serum as a hypothermic solution may be appropriate. However, plasma from donors who are not healthy is associated with risk. Residual drugs and chemotherapeutic agents may be present. Donors who are ill may also have a poor nutritional status, which, in turn, may also influence the stability of cells stored in plasma. Using plasma represents a source of variability that could have an adverse effect on the stability of cells during hypothermic storage.

For cells in culture, there are a limited number of commercially available hypothermic storage solutions. Tissue culture media is usually not suitable for use as a short-term storage solution, as most buffering systems require the use of a 5% CO_2 culture environment (Freshney 2010). Loss of buffering results in significant shifts in pH and may result in stress for the cells that can accelerate losses. Organ storage solutions are very expensive and are formulated to address damage to organs resulting from ischemia and reperfusion injury (versus damage due to reactive oxidative species).

As described in Chapter 1, umbilical cord blood (UCB) is a common cell type that is subjected to short-term storage prior to cryopreservation. UCB is collected any time of the day or night and is typically depleted of red blood cells (RBCs), followed by cryopreservation and storage in a UCB repository. Processing of the UCB units is performed in specialized facilities (i.e., Good Manufacturing Practices (GMP) cell processing facilities) and not in the hospital where the UCB is collected. Therefore, it must be shipped, typically in the liquid state, from the site of collection to the site of processing and storage. The cells are shipped as collected and no additional storage solutions or additives (other than anticoagulants) are added. In this context, the baby's own blood plasma acts as a short-term storage solution.

There are a limited number of short-term storage solutions that have been developed for liquid storage of cells. As mentioned previously, RBCs are stored in solutions designed specifically for RBCs (Valeri et al. 2000) and cannot be used for other cell types. A short-term storage solution for hematopoietic cells was developed and tested on UCB, bone marrow (Schmid et al. 2002), and peripheral blood stem cells (Burger, Hubel, and McCullough 1999).

Improved methods for hypothermic storage are needed that enable storage of cells for duration beyond 72 h. The ability to extend nonfrozen storage of cells for up to a week could be transformative for the field of regenerative medicine and cell therapy. A recent study suggests that physical stabilization (preventing settling of the cells) can be used to improve recovery of cells stored hypothermically (Wong et al. 2016). This methodology will be important in extending the duration of storage.

Selection of Subpopulations

It is common for cell populations to be heterogeneous. An apheresis product that is used as the basis for a cell therapy product will contain a variety of cell types. This heterogeneity can influence the behavior of the mixed populations. T-cells can be isolated from the apheresis product. The presence of regulatory T-cell subpopulations present in the isolated T-cells can influence the behavior of the product given therapeutically. In a population of cells that have been genetically modified, genetic modification does not occur in all of the cells; therefore, some of the cells may be modified and others not. The behavior of the product may vary with varying composition (genetically modified versus non-modified cells). Therefore, purification of cells or removal of undesirable cell types is a common process. Two of the most common methods of cell selection include immunomagnetic isolation and flow sorting of cells.

Immunomagnetic methods involve labeling of cells with magnetic nanoparticles conjugated to specific antibodies. The labeled cells are passed through a magnet and the cells that are labeled are retained by the magnetic field, while the unlabeled cells pass through. If the labeled cells are used, the process is known as enrichment/selection. If the labeled cells are discarded, the process is known as depletion (i.e., removal of an unwanted cell type). The process of cell selection can involve multiple washings of cells and take several hours, which can stress the cells.

Another method of selecting subpopulations of cells involves the sorting of cells using flow cytometry. The cells are labeled with fluorophores conjugated typically to antibodies that label target cells, which are then physically sorted through the device based on the cell surface fluorescence. As with immunomagnetic separation, the process can result in either enrichment or depletion of a given population. The process can take a considerable period of time and processing of the sample to achieve the desired subpopulations.

Activation or Stimulation

Certain processes act to change the biological state or properties of a cell without modification of a cell's genome. The most common method by which this is accomplished is activation or stimulation. Activation is commonly used with immune cell-based therapies. Specifically, T-cells can be activated through coculture with antigen expressing cells or an artificial construct such as nanoparticles conjugated with antigen. The end result is activation of the cell, which influences a variety of biological properties such as proliferation, cell metabolism, and function and resistance to apoptosis (Palmer et al. 2015). There have been limited studies of the freezing behavior of activated cells. One study suggests that activated T-cells are more resistant to post-thaw apoptosis (Hubel 2006). This study suggests that it may be possible to influence response to the stresses of freezing and thawing by altering the biology of cells through nongenetic methods.

Genetic Modification

A variety of methods can be used to genetically modify cells. A common method for genetic modification is a viral vector, which is a recombinant virus designed to deliver genetic material into cells. In nature, viruses infect cells and deliver genetic material into cell. Viral vectors utilize those same mechanisms to transfer genetic material into cells. Different types of viral vectors have been used to genetically modify cells including retroviruses, lentiviruses, adenoviruses, and adeno-associated viruses. The type of viral vector can be selected based on specific biological attributes. For example, lentiviruses may be used in applications where the target cells are nonproliferating, whereas retroviruses require dividing cells for integration.

Recent developments in gene-editing technology have enabled insertion, deletion, or replacement of genes inside a cells. Engineered nucleases are used to create site-specific double-strand breaks at desired locations, and the breaks are repaired through either homologous recombination or nonhomologous end-joining. Meganucleases, zinc finger nucleases, transcription activator-like effector-based nucleases, and Clustered Regularly Interspaced Short Palidromic Repeats (CRISPR-Cas 9) are methods of gene editing currently being used. CRISPR-Cas 9, for example, can involve the use of both viral and nonviral methods (e.g., electroporation, ribonucleocomplexes) for insertion of the genes.

The addition or deletion of gene(s) may influence the freezing response of cells. There is little scientific literature examining this issue. The differences in freezing responses of different cells suggest that differences in gene expression may influence freezing response. In addition, there are certain patient-specific induced pluripotent stem (iPS) cells that do not respond well to conventional cryopreservation protocols. As the use of gene-editing technologies grows, certain genetic modifications may require modifications of the cryopreservation protocols.

Culture

It is common for both primary cells and cell lines to be cultured prior to cryopreservation. As with other forms of pre-freeze processing, culture conditions can influence cell health and therefore the ability of the cells to survive the additional stresses of freezing and thawing. Culture conditions should avoid nutrient deprivation or excessive shear (for hollow fiber bioreactors or fluidized bed reactors).

The most common reason to perform cell culture is to expand cell number. For primary cells, cells should be frozen at a relatively low passage number (and the passage number should be part of the annotation of the sample). Excessive culture of primary cells can lead to shifts in phenotype (Freshney 2010). For mesenchymal stromal cells (MSCs) expanded in culture, a shift in phenotype from normal to senescent was observed within a relatively small number of passages (approximately 5). The senescent phenotype survived freezing but did not persist (Pollock et al. 2015). Unfortunately, the senescent phenotype is an inflammatory one and not desired for clinical application of the cell. For continuous cell lines, it is recommended that cells being cryopreserved should be in exponential growth phase (Coecke et al. 2005) in order to promote a high ratio of nucleus to cytoplasm.

Pre-freeze Process Monitoring

Stressing of the cells due to pre-freeze processing may not necessarily produce a loss of viability (i.e., loss of membrane integrity). The cells can be monitored during processing for early signs of apoptosis or shifts in metabolism as markers for sublethal stressing of the cells.

Pre-freeze Characterization

Proper downstream use of cryopreserved cells requires knowledge of the cell type that has been preserved. Improper cell identification is a common source of error in biomedical research (Freedman et al. 2015) As a result, cell characterization is a critical step in preservation processing. Pre-freeze characterization may include determination of the following:

- Identity
- Genetic stability
- Enumeration
- Purity
- Adventitious infectious agents

It is noteworthy that each method of cell characterization has limitations. Most systems are accurate for a given cell size or range of cell concentrations. Knowledge of the operating characteristics of the specific method of cell characterization is critical to proper use.

Identity

Cell lines: Determining the identity of cell lines prior to cryopreservation is critical as studies have demonstrated that a significant number of cells lines used in biomedical research are misidentified or contaminated with other cell types (American Type Culture Collection Standards Development Organization Workgroup 2010). The prevalence of cell line misidentification is so common that funding agencies are now requiring routine identification of cell lines used for biomedical research. Fortunately, the commercial availability of cell identification assays based on single nucleotide polymorphisms or short tandem repeats makes identification of cells lines rapid and relatively inexpensive.

Primary cells: Primary cells are isolated from tissues, organs, or circulating blood. In contrast to cell lines, primary cells are identified most often through cell surface markers. Cells are stained with fluorescent antibodies and the cells are interrogated using flow cytometry for expression (or lack thereof) of specific surface markers. Cells of a hematopoietic lineage express $CD45^+$. CD31 is normally found on endothelial cells, platelets, macrophages, Kupffer cells, granulocytes, lymphocytes (T-cells, B cells, and NK cells), megakaryocytes, and osteoclasts. Epithelial cell adhesion molecule is expressed exclusively in epithelia and epithelial-derived neoplasms. As certain surface markers are expressed on a variety of cell types, it is common for cells to be identified using a combination of surface markers. For example, MSCs are identified using the combination of cell surface markers $CD73^+CD90^+CD105^+CD45^-$ (Dominici et al. 2006).

Hematopoietic stem cell products (e.g., UCB, peripheral blood progenitor cells, and bone marrow) are cells commonly used therapeutically. For allogeneic cells (cells taken from one donor and given to another), characterization will include Human Lymphocyte Antigen and ABO compatibility in order to enable proper matching of cells to the recipient. The hematopoietic progenitor cell ($CD34^+CD45^+$) subpopulation is capable of reconstituting hematopoiesis in a recipient and characterization of HSC products also includes quantification of this population of cells, typically by flow cytometry.

If heterogeneous cultures are isolated from primary tissue, the characteristics of the culture may vary with time. For example, tumor explants may contain epithelial cells associated with the specific type of cancer. Stromal cells in the tumor are also present and may overgrow the culture with time as they proliferate more rapidly than the epithelial cells. As a result, characterization of heterogeneous cultures may need to be performed on a regular basis.

Genetic Stability

Mutation, mismatch repair deficiency, and chromosomal instability are three principal types of genetic instabilities. Cells that are cultured extensively and/or have been reprogrammed should be analyzed for genetic stability.

Induced pluripotent stem cells (iPS cells) and embryonic stem cells (ESC): Both iPS cells and ESCs can exhibit genetic instability with passage in culture, in particular, there is a tendency to acquire changes affecting chromosomes 1, 12, 17, and 20. Karyotyping and fluorescence in-situ hybridization (FISH) are two techniques used commonly to verify the genetic stability of these cells (Steinemann, Gohring, and Schlegelberger 2013).

Cell lines: Cell lines that are genetically stable are required for the production of proteins that are biochemically comparable throughout the production cycle. Loss of genomic DNA or overgrowth of a specific altered clone can result in the production of an abnormal glycosylation pattern for the protein or posttranslation modifications, which in turn may influence the therapeutic potential of the protein. Determining the genetic stability of a cell line varies with application and cell type. In general, however, the transgene DNA and the number of transgene copies and subsequent mRNA sequence for a given cell line must remain stable over the length of time required for the production of a protein.

Enumeration

Counting the total number of cells present in a sample is a very common method of characterizing cells, especially for cells used therapeutically. The most common method for enumerating cells using a manual method is with a hemocytometer (or similar reticulated grid). This device is easy to use and requires minimal training. However, there are common sources of error for the use of the device including (i) the sample must be mixed uniformly prior to loading in the chamber, (ii) proper dilution of the sample (approximately 1–2 million cells/ml), (iii) counting a sufficient number of events (200–300 cells), and (iv) evaporation of the sample. This method of counting is used most commonly for laboratory enumeration or when the sample does not permit use of automated counters.

A common method of quantifying cells in aqueous solution involves counting changes in electrical impedance when the cells pass through an aperture (Creer 2016). The change in the impedance is a function of the cell volume, so it is also a method of quantifying cell size. Differential counts can be performed using this method but there will be some bias to those measurements. For example, nucleated RBCs will be counted as lymphocytes, which is a problem when processing UCB but not other hematopoietic stem cell products. Cell enumeration using radio frequency conductance is also available and operates in a similar manner to electrical impedance.

The second most common method of automated enumeration of cells involves using light (e.g., laser, fluorescent, or conventional light). Forward and side scattering of the light can correlate with the size as well as the complexity of the cells. Flow cytometry uses this method for cell analysis, and there are other commercially available systems that also use light.

Purity

Peripheral blood or hematopoietic stem cell products are heterogeneous cell products, and clinical analyzers described above will produce cell counts and differentials (percentage of respective cell types present in the mixture) for these products. The differential cell counts is used clinically to detect health or disease and may also be used for characterizing cell products during specific processing steps, in particular selection of subpopulations.

The progenitor cell population (CD34$^+$) present in hematopoietic stem cell products is a clinically important but relatively small percentage (2–4%) of peripheral blood from mobilized donors, bone marrow, and UCB. Considerable work has been performed to standardize the enumeration of this cell type (Sutherland et al. 1996). The standardization of the acquisition and analysis of this cell type using flow cytometry was intended to reduce variation between both operators and laboratories.

Adventitious Agents

Cells infected with endogenous agents should not be cryopreserved and best practices suggest that cells should be tested prior to cryopreservation. Specific testing regimes will depend on the donor source, the culture history, and their intended use. For blood products, current testing includes hepatitis B, hepatitis C, human immunodeficiency virus (Types 1 and 2), human T-lymphotrophic virus (HTL-I/II), syphilis, West Nile, and anti-*Trypanosoma cruzi*. Organ donors may be screened for Epstein Barr Virus and cytomegalovirus in addition to the diseases listed previously.

Pre-freeze processing may also result in contamination of the cells with adventitious agents, which include viruses, bacteria, mycoplasma, fungi, rickettsia, protozoa, parasites, and transmissible spongiform encephalitis agents. Cells should be screened for the presence of these contaminants. Commercial testing laboratories have standard batteries of tests designed to screen for these agents. Additional testing may be necessary, however, depending upon the processing of the cells. For example, if cells have been cultured with media containing bovine or fetal calf serum, it may be appropriate to have the cells tested for bovine viruses.

Microbial Testing of Cell Therapy Products

Cell therapy products most often consist of a batch size of one, and sterility testing is commonly a release criteria. Therefore, the sample volume required for each test may represent a significant volume or cell content. As a result, the amount of material used for microbial testing is typically small (less than 1 ml).

The sampling of the product should occur after the sample is in the final container and after all solutions have been added. A risk-based approach is sometimes used for determining whether or not post-thaw testing is appropriate. For example, post-thaw testing may be appropriate if there was cracking or failure of the container during freezing, storage, or thawing. The shelf-life of a frozen and thawed cell therapy product is typically less than the time required for completion of sterility testing (Duguid, Khuu, and du Moulin 2016).

New methods of assessing contamination are being developed and these methods are available commercially. Several of these methods could be applicable for use in pre-freeze testing of cells:

- Bioluminescent assays in which the release of ATP from microorganisms is detected
- Changes in the electrical impedance of a solution resulting from by-products of metabolism produced by bacterial growth
- Nucleic acid–based methods capable of detecting different microorganisms (e.g., PCR)
- Detection of changes in gas pressure resulting from bacterial growth
- Flow cytometry combined with fluorescent stains to detect bacterial cells

These methods may detect contamination quickly versus conventional methods that may require days.

Special Considerations for the Characterization of Cell Therapies

The identity of cells used therapeutically is handled differently and reflects regulatory requirements (21 CFR Part 210, 211, 312, and 1271). The required testing involves determining safety, purity, identity, potency, and stability of the cells prior to freezing. Purity of the cell product is typically determined by cell counts and cell surface marker characterization as described previously. Identity is determined typically through quantification of cell surface markers. Potency can be determined by viability and functionality assays, which may vary with cell type and application. As with banked cells, cells used therapeutically will be tested for human viruses and for fungal, bacterial, and mycoplasm contamination. Sterility testing for most products should be performed in agreement with Section 361 of the Public Health Service (PHS) Act (42 USC 264) or Section 351(a) of the PHS Act (42 USC 262(a)) depending upon the product.

Annotation of Pre-freeze Processing

Best practices require that the processing record of the sample should be annotated to reflect all of the processing steps prior to cryopreservation. For blood-based biospecimens, the annotation of the sample has been standardized. Standard Preanalytical Codes (SPREC) (Betsou et al. 2010) is an annotation system that has been developed by the International Society for Biological and Environmental Repositories (ISBER) for fluid biospecimens. The factors that influence post-thaw recovery of cells from blood and other fluid biospecimens include the following:

- The source of the cells (blood, urine, ascites, bronchial alveolar lavage, etc.)
- Type of container
- Anticoagulants or other additives in the collection tube (protease inhibitors)
- Delay time between collection and processing
- Temperature at which the sample was held during the delay
- Centrifugation speed
- Post-centrifugation delay

Cell therapies based on hematopoietic cells may be obtained from apheresis or bone marrow aspiration. For bone marrow aspiration, annotation of the sample may include the following:

- Anesthesia used
- Collection technique (location, etc.)
- Anticoagulant
- Volume collected

- Number of nucleated cells harvested
- Time and temperature between collection and processing
- Centrifugation (duration and g-force used)
- Bag and anticoagulant
- Centrifugation
- Filtration
- Separation of mononuclear cells

For apheresis collection, annotation may include the following:

- Regime for mobilization (if used)
- Apheresis device used
- Operating conditions for device (i.e., inlet flow rate)
- Cell counts (RBC contamination, total cell count, etc.)
- Collection efficiency
- Anticoagulant and bag used for collection
- Time and temperature between collection and processing

Additional information on factors that can influence pre-freezing processing for common cell therapy processes can be found in Areman and Loper (2016).

The lists of pre-analytical factors given previously is not meant to be exhaustive, but suggestive of the nature of the factors that influence the quality of the cells prior to cryopreservation. Proper annotation of the sample includes a record of these parameters, *even if they were not specifically controlled during processing*. This approach will permit continuous improvement in the collection and processing of samples.

Scientific Principles

- *The processing of cells prior to cryopreservation can influence the response of the cells to the freezing process.*

Putting Principles into Action

- Samples should be annotated and the pre-freeze processing that the sample has been subjected to should be noted as a part of the sample record using SPREC or other appropriate guidelines (Betsou et al. 2010).
- It is important to characterize the cells before they are cryopreserved. Identity of the cells should be determined and the cells should be tested for endogenous or adventitious agents prior to cryopreservation.
- The process is the product, so starting with healthy cells is a critical step to a favorable outcome post-thaw.

- Determination of membrane integrity using dyes is one method of detecting stress. Viabilities < 80% may indicate problems with the processing that result in pre-freeze stressing of the cells.
- Some stresses associated with pre-freeze processing are sublethal but may influence post-thaw recovery. Checking for early signs of apoptosis (caspase expression) may also be helpful. High levels of caspase expression may indicate pre-freeze processing is stressful.
- If the cells are cultured, checking for shifts in phenotype may be helpful. Specifically, quantifying the fraction of cells that are senescent may help characterize the quality of the cultures. The fraction of cells that are senescent should be less than 5%.
- When cells have been stressed, certain strategies can be used to improve the health of the cells prior to cryopreservation.
- For cells digested from organs or tissues, brief periods of culture pre-freeze may improve post-thaw recovery of cells.
- For anchorage-dependent cells, culture of cells in spheroids has also been used to reduce anoikis and improve post-thaw recovery of hepatocytes (Darr and Hubel 2001, Hubel and Darr 2004).
- The addition of caspase inhibitors or Rho-associated kinease protein inhibitor (ROCK inhibitor) can be used to inhibit apoptosis. However, the use of caspase inhibitors or ROCK inhibitors is associated with reduced proliferation post-thaw.

References

American Type Culture Collection Standards Development Organization Workgroup, A. S. N. 2010. "Cell line misidentification: the beginning of the end." *Nat Rev Cancer* 10 (6):441–448.

Areman, E.M., and K Loper. 2016. Cellular Therapy: Principles, Methods and Regulations. 2nd ed. Bethesda, MD: AABB.

Betsou, F., S. Lehmann, G. Ashton, M. Barnes, E. E. Benson, D. Coppola, Y. DeSouza, J. Eliason, B. Glazer, F. Guadagni, K. Harding, D. J. Horsfall, C. Kleeberger, U. Nanni, A. Prasad, K. Shea, A. Skubitz, S. Somiari, E. Gunter, Biological International Society for, and Science Environmental Repositories Working Group on Biospecimen. 2010. "Standard preanalytical coding for biospecimens: defining the sample PREanalytical code." *Cancer Epidemiol Biomarkers Prev* 19 (4):1004–1011.

Burger, S. R., A. Hubel, and J. McCullough. 1999. "Development of an infusible-grade solution for non-cryopreserved hematopoietic cell storage." *Cytotherapy* 1:123–133.

Coecke, S., M. Balls, G. Bowe, J. Davis, G. Gstraunthaler, T. Hartung, R. Hay, O. W. Merten, A. Price, L. Schechtman, G. Stacey, W. Stokes, and Ecvam Task Force on Good Cell Culture Practice Second. 2005. "Guidance on good cell culture practice. a report of the second ECVAM task force on good cell culture practice." *Altern Lab Anim* 33 (3):261–287.

Creer, M.H. 2016. "Integrated Analysis of Hematopoietic Cellular Therapy Product Quality." In Cellular Therapy: Principles, Methods and Regulations, edited by E.M. Areman and K. Loper, 530–544. Bethesda, MN: AABB.

Darr, T. B., and A. Hubel. 2001. "Postthaw viability of precultured hepatocytes." *Cryobiology* 42 (1):11–20.

Dominici, M., K. Le Blanc, I. Mueller, I. Slaper-Cortenbach, F. Marini, D. Krause, R. Deans, A. Keating, Dj Prockop, and E. Horwitz. 2006. "Minimal criteria for defining multipotent mesenchymal stromal cells. The International Society for Cellular Therapy position statement." *Cytotherapy* 8 (4):315–317.

Duguid, J., H. Khuu, and G.C. du Moulin. 2016. "Assessing Cellular Therapy Products for Microbial Contamination." In Cellular Therapy: Principles, Methods and Regulations, edited by E.M. Areman and K. Loper, 606–611. Bethesda, MD: AABB.

Freedman, L. P., M. C. Gibson, S. P. Ethier, H. R. Soule, R. M. Neve, and Y. A. Reid. 2015. "Reproducibility: changing the policies and culture of cell line authentication." *Nat Methods* 12 (6):493–497.

Freshney, R. I. 2010. Culture of Animal Cells: A Manual of Basic Technique and Specialized Applications. 6th ed. Hoboken, NJ: Wiley-Blackwell.

Hubel, A. 2006. "Cellular Preservation: Gene Therapy, Cellular Metabolic Engineering." In Advances in Biopreservation, edited by J. G. Baust. Boca Raton, FL: CRC Press.

Hubel, A, and T. B. Darr. 2004. "Post-thaw function and caspase activity of cryopreserved hepatocyte aggregates." *Cell Preserv Tech* 2 (3):164–171.

Hubel, A., M. Conroy, and T. B. Darr. 2000. "Influence of preculture on the prefreeze and postthaw characteristics of hepatocytes." *Biotechnol Bioeng* 71 (3):173–183.

Li, A. P., P. D. Gorycki, J. G. Hengstler, G. L. Kedderis, H. G. Koebe, R. Rahmani, G. de Sousas, J. M. Silva, and P. Skett. 1999. "Present status of the application of cryopreserved hepatocytes in the evaluation of xenobiotics: consensus of an international expert panel." *Chem Biol Interact* 121 (1):117–123.

Nyberg, S. L., J. A. Hardin, L. E. Matos, D. J. Rivera, S. P. Misra, and G. J. Gores. 2000. "Cytoprotective influence of ZVAD-fmk and glycine on gel-entrapped rat hepatocytes in a bioartificial liver." *Surgery* 127 (4):447–455.

Palmer, C. S., M. Ostrowski, B. Balderson, N. Christian, and S. M. Crowe. 2015. "Glucose metabolism regulates T cell activation, differentiation, and functions." *Front Immunol* 6:1–6.

Pollock, K., D. Sumstad, D. Kadidlo, D. H. McKenna, and A. Hubel. 2015. "Clinical mesenchymal stromal cell products undergo functional changes in response to freezing." *Cytotherapy* 17 (1):38–45.

Rauen, U., and H. de Groot. 2002. "Mammalian cell injury induced by hypothermia—the emerging role for reactive oxygen species." *Biol Chem* 383 (3–4):477–488.

Schmid, J, J. McCullough, S. R. Burger, and A. Hubel. 2002. "Non-cryopreserved bone marrow storage in STM-Sav, in infusible-grade cell storage solution." *Cell Preserv Technol* 1 (1):45–52.

Steinemann, D., G. Gohring, and B. Schlegelberger. 2013. "Genetic instability of modified stem cells—a first step towards malignant transformation?" *Am J Stem Cells* 2 (1):39–51.

Sutherland, D. R., L. Anderson, M. Keeney, R. Nayar, and I. Chin-Yee. 1996. "The ISHAGE guidelines for CD34+ cell determination by flow cytometry. International Society of Hematotherapy and Graft Engineering." *J Hematother* 5 (3):213–226.

Valeri, C. R., L. E. Pivacek, G. P. Cassidy, and G. Ragno. 2000. "The survival, function, and hemolysis of human RBCs stored at 4 degrees C in additive solution (AS-1, AS-3, or AS-5) for 42 days and then biochemically modified, frozen, thawed, washed, and stored at 4 degrees C in sodium chloride and glucose solution for 24 hours." *Transfusion* 40 (11):1341–1345.

Wong, K. H., R. D. Sandlin, T. R. Carey, K. L. Miller, A. T. Shank, R. Oklu, S. Maheswaran, D. A. Haber, D. Irimia, S. L. Stott, and M. Toner. 2016. "The role of physical stabilization in whole blood preservation." *Sci Rep* 6:1–9.

Yagi, T., J. A. Hardin, Y. M. Valenzuela, H. Miyoshi, G. J. Gores, and S. L. Nyberg. 2001. "Caspase inhibition reduces apoptotic death of cryopreserved porcine hepatocytes." *Hepatology* 33 (6):1432–1440.

3

Formulation and Introduction of Cryopreservation Solutions

Importance of Cryoprotective Agents

Attempts to successfully freeze biological samples have been pursued since the early 1800s (McGrath et al. 1988). The ability to cryopreserve biological samples and regain viability and cell function upon thawing was not realized until the discovery of agents that provide protection against the stresses of freezing and thawing. These agents are more commonly known as cryoprotective agents (CPAs). The first CPA to be discovered was glycerol in 1949 (Polge, Smith, and Parkes 1949), followed roughly 10 years later by the discovery of dimethylsulfoxide (DMSO) and its cryoprotective capabilities (Lovelock and Bishop 1959). These two CPAs are used widely for the formulation of cryopreservation solutions.

In a very general sense, CPAs are additives that alter the behavior of water during freezing and suppress ice formation. These molecules must also stabilize proteins, the cell membrane, and other critical structures in the cell. Traditionally, the search for new or improved CPAs has relied on finding or synthesizing a single molecule capable of providing all of these modes of action. DMSO has been that molecule for several decades. DMSO has significant limitations (including cell toxicity and the fact that it is ineffective for certain cell types), thus, alternatives to DMSO for stabilizing cells has been an active area of research. Alternatives to DMSO may involve development of optimized solutions containing mixtures of agents capable of acting in concert to preserve cells. As our understanding of the sites and modes of action for CPAs continues to evolve, the potential for customized design of those molecules or *in silico* screening of molecules for their cryoprotective properties becomes possible. The end result will be the development of a larger library of compounds that can protect cells and potentially the ability to customize cryopreservation solutions for specific applications or specific cell types.

Other molecules have been shown to provide a cryoprotective benefit but are less widely used. A listing of molecules that have been shown to provide cryoprotective benefits can be found in Table 3.1. A limited number of molecules

Preservation of Cells: A Practical Manual, First Edition. Allison Hubel.
© 2018 John Wiley & Sons, Inc. Published 2018 by John Wiley & Sons, Inc.

Table 3.1 Cryoprotective molecules.

Molecules shown to provide a cryoprotective benefit to a wide variety of biological systems

Dextran	Polyvinylpyrrolidone
Dimethylsulfoxide	Propylene glycol
Ethylene glycol	Trehalose
Glycerol	Sucrose
Hydroxyethyl starch	

Molecules that have shown a cryoprotective benefit in a more limited context

Alanine	Mannose
Albumin	Methanol
Butandiol	Methoxy propanediol
Chondroitin sulfate	Methyl acetamide
Choline	Methyl formamide
Diethylene glycol	Methyl glucose
Dimethyl acetamide	Methyl glycerol
Dimethyl formamide	3-O-Methyl-D-glucopyranose
Erythritol	Proline
Formamide	Propandiol
Glucose	Ribose
Glycerolphosphate	Serine
Glycerolmonoacetate	Sorbitol
Glycine	Triethylene glycol
Inositol	Trimethylamine acetate
Lactose	Urea
Maltose	Valine
Mannitol	Xylose

(such as DMSO and glycerol) provide a cryoprotective benefit for a wide variety of biological samples, whereas other molecules exhibit cryoprotective benefits for a limited number of cells. It is noteworthy that a variety of molecules have demonstrated cryoprotective benefits: organic solvents (e.g., DMSO, methanol), polymers (e.g., detran, hydroxyethly starch), sugar alcohols (e.g., glycerol, sorbitol, mannitol, inositol), sugars (e.g., sucrose, glucose, trehalose) and organic compounds (including amino acids).

Mechanisms of Cryoprotection

Historically, modes of action for CPAs were framed in terms of their action on water and its behavior during freezing. This emphasis is not surprising as water plays an important role in biological systems. The phase diagram for common multicomponent solutions used in cryopreservation has been characterized (see Cocks and Brower 1974). The addition of solute changes the concentration of salts at a given subzero temperature and this effect is commonly known as the colligative effect. Other studies address the effect of CPAs on the formation of glasses (i.e., vitrification) of aqueous solutions (Angell 1995) both outside and inside the cell. Recent studies provide insight into the molecular-level behavior of water during freezing, in particular, for complex solutions used in cryopreservation. Spectroscopic studies (Vanderkooi, Dashnau, and Zelent 2005) have demonstrated that CPAs such as glycerol or sucrose influence the hydrogen-bonding behavior of water.

Since cells contain more than just water, the post-thaw function of the cell requires integrity of the cell membrane and function of intracellular components (e.g., cytoskeleton, proteins, nucleus). Studies have also demonstrated the role of cryoprotectants on stabilizing critical biological structures of the cell. Sugars, specifically trehalose, stabilize the cell membrane during freezing (Anchordoguy et al. 1987). Other osmolytes stabilize proteins (Arakawa and Timasheff 1985). Combinations of osmolytes preserve post-thaw functions of cells (Pollock et al. 2016b) and stabilize the cytoskeleton (Pollock et al. 2017).

Formulating a Cryopreservation Solution

Cryopreservation medium can be formulated in-house or premanufactured formulations can be purchased from a variety of sources and come with well-characterized components and known storage and stability. Many of the formulations available commercially are proprietary, which has been a cause for concern amongst users.

Whether using preformulated solutions or solutions formulated in-house, the quality of reagents used in formulating a cryopreservation solution is critical. Cells suspended in a cryopreservation solution cannot be sterilized, filtered, or purified. Therefore, the quality of the reagents determines the quality of the cryopreservation solution. Ideally, components of cryopreservation solutions should be of pharmaceutical grade. Residual contaminants resulting from the manufacturing process have been found to influence post-thaw recovery of cells for a variety of commonly used CPAs, so care should be used to source high-quality, well-defined components in the freezing solution (Sputtek 1991).

Table 3.2 Common cryopreservation solution compositions for *in vitro*, cell therapy, and cell-banking applications.

In vitro applications	Cell therapy applications	Cell-banking applications
Tissue culture medium	Solution approved for human infusion such as	Defined tissue culture medium
Cryoprotective agents	• Normosol R	Cryoprotective agents
Animal serum	• Plasmalyte A	
	• Lactated ringers	
	Cryoprotective agents	
	Human serum albumin	

Most cryopreservation solutions have three basic components: a carrier solution, cryoprotective agent(s) (CPAs), and a source of protein. The carrier solution is the principal component by volume and is typically a balanced salt solution. Cryopreservation solutions vary depending upon the application (see Table 3.2).

For in vitro/research applications, it is common for the carrier solution of a cryopreservation solution to use tissue culture medium. In certain applications, a protein source (i.e., plasma or serum) has been used as the carrier solution of a cryopreservation solution.

For cell therapy/in vivo applications, an infusible grade-balanced salt solution like Normosol-R or Plasmalyte A is commonly used as the carrier solution of a cryopreservation solution. Tissue culture medium contains components that are not approved for human infusion and may contain additives that are known pyrogens. Fetal bovine serum (or other animal proteins are not suitable for use in cryopreservation solutions for human therapeutic applications. Animal proteins will elicit an immune response from patients when injected (in particular for therapies requiring multiple doses) and there is the potential for transmission of adventitious agents (e.g., bovine spongiform encephalitis). Human serum (HS), platelet lysate, or plasma may also be used in cryopreservation solutions. These products are not defined, and composition or effect on preservation outcome can vary with health status of the donor and, for patients, may contain residual drugs or chemotherapeutic agents. One alternative that has been used is HS albumin, which is a more consistent product but still has issues in terms of potential for disease transmission.

In cell-banking applications, preservation of cell lines used in the production of biological products typically use culture medium (defined without serum) supplemented with CPAs. It is common for cell-banking applications to avoid using animal products in cryopreservation solutions. Serum is not a defined component and can vary from lot to lot with concerns over disease transmission as well. As a result, it is not commonly used in cell-banking applications.

Other additives to cryopreservation solutions include the following:

- Anticoagulants and DNAses. Cells that lyse during freezing can result in aggregation of cells or clotting of proteins present in the residual plasma. In order to reduce or prevent aggregation, anticoagulants or DNAse are frequently added to reduce coagulation of the sample or aggregation of cells.
- Buffering agents. Under atmospheric conditions, tissue culture medium is not buffered and will experience significant shifts in pH. Similarly, electrolyte solutions are not buffered. Certain cell types are more sensitive to pH shifts than others and pre-freeze stressing of the cells due to pH shifts may have an adverse influence on post-thaw recovery. Cryopreservation solutions may include a buffering agent appropriate for use under normal atmospheric conditions. After formation of ice in the extracellular solution, shifts in pH are expected to be independent of the buffering system used (Bhatnagar, Bogner, and Pikal 2007).
- Biological modifiers, which include caspase inhibitors or other anti-apoptosis agents and Rho-associated kinease protein inhibitors (ROCK). The stress of pre-freeze processing can result in apoptosis or cell death. ROCK inhibitors have been used with induced pluripotent stem cells to improve post-thaw recovery. Similarly, the stresses of freezing and thawing can result in post-thaw apoptosis; therefore, caspase inhibitors have been used to reduce post-thaw cell losses.

Formulation of a Vitrification Solution

Pure water can be vitrified but only under special conditions (small volumes and very rapid cooling rates). Most vitrification solutions currently in use for preservation protocols consist of a carrier solution and high concentrations of CPAs (4–6 M is common) (see Fahy and Wowk 2015, for review). Traditionally, CPAs commonly used in vitrification solutions include DMSO, glycerol, ethylene glycol, propylene glycol, sucrose, and polymers. The concentration and combination of CPAs are designed to suppress the formation of ice during freezing.

Other components in vitrification solutions may include agents designed to block nucleation of ice or growth of ice crystals (i.e., ice blockers). Antifreeze proteins act to inhibit growth of ice, but they can be difficult to obtain or expensive for the levels required. Other options include polymers such as polyvinyl alcohol or polyglycerol, which have been shown to be effective. Ice blockers may also reduce the total concentration of solutes required for vitrification of the sample.

As toxicity of vitrification solutions is an important limitation in their use, vitrification solutions may also include additives to reduce the kinetics of damage after introduction (see Fahy 2010, for review) of the solution. For example, amides are used to reduce the toxicity of DMSO. Specifically, formamide and urea can be used to reduce toxicity. Other additives such as acetamide and N-methylacetamide are also used but may be less effective.

Characterization and Quality Control for Cryoprotective Solutions

The use of incorrect or expired cryopreservation media has been identified as sources of variation in freezing outcome (Freedman et al. 2015). Therefore, cryopreservation solutions formulated in-house should be subjected to reasonable quality control measures. For example, the solution pH and osmolarity should be measured after formulation. An osmometer is not expensive and is easy to use. This device can be used to quantify the total osmolarity of the solution. CPAs, salts, and other additives may contribute to the overall osmolarity of the solution. The solution osmolarity could be compared against what was expected and ranges of acceptable osmolarities specified. This practice will reduce errors in formulation or dilution of cryopreservation solutions.

As the shelf-life of a cryopreservation solution is also critical, additional stability studies should be performed to establish the shelf-life for a formulated cryopreservation solution. Variations in the pH of the solution with time are an indication of degradation of components. Fourier Transform Infrared Spectroscopy can also be used to characterize the stability of a formulated cryopreservation solution. Spectra of the solution can be obtained and monitored as a function of time. Changes in the wavenumber or peak size of the spectra can identify degradation of the solution and the specific components that are degrading.

An additional approach would be to measure post-thaw recovery of a commonly used cell line as a function of time after formulation of the solution. Significant drops in the post-thaw recovery of a given cell line with age of cryopreservation solution could be used to establish the shelf-life for that solution. Once a shelf-life has been established, protocols should specify shelf-life of the solution and disposal of out-of-date solutions.

Basic quality control measures on the cryopreservation solutions will help improve the reproducibility of freezing protocols. Many of the measures described above are not difficult or expensive, yet should be performed in order to contribute to the reproducibility of preservation processes.

In addition to the characterization studies described earlier, the freezing behavior of vitrification solutions may also be characterized using differential scanning calorimetry. Heat release associated with ice formation can be detected, and these studies are commonly used to verify that the solution as formulated does *not* form ice.

Toxicity of CPAs

Cryopreservation solutions are not physiological solutions, and most cryo-preservation solutions are hypertonic. The solutions are designed to protect the cells during freezing. In general, leaving the cells in a cryopreservation solution for long term is not desirable. Osmotic and biochemical toxicities are two independent mechanisms of damage that can occur during introduction, incubation, and removal of a cryopreservation solution. Proper development of a cryopreservation protocol requires characterization of those mechanisms of damage for a given cell type and solution composition.

Osmotic Toxicity

The osmolarity of a physiological solution is 270–300 mOsm. In contrast, a 10% DMSO solution has an osmolarity of approximately 1400 mOsm. When cells are removed from a physiological solution and placed in a cryopreservation solution, the initial response of the cells will be to dehydrate (lose water) in order to reduce the chemical potential difference between the intracellular and the extra-cellular solutions. If the solution contains small molecules (e.g., DMSO and glyc-erol), the cells will dehydrate rapidly followed by a slow increase in cell volume, resulting from penetration of the small molecule into the cell (Figure 3.1a). The molecular weight of water is 18 Da and, in contrast, DMSO is 78 Da and glycerol is 92 Da. These differences in molecular weights imply that water will always move more rapidly into and out of the cell, resulting in significant changes in cell volume. If these volumetric excursions are large enough, the cells will lyse upon introduction of the cryopreservation solution. If the cryopreservation solution contains only large molecules that exert an osmotic force but do not passively penetrate the cell membrane, the cells will dehydrate and remain dehydrated (Figure 3.1b). As osmotic sensitivity can vary amongst different cell types, easy methods of determining sensitivity of a given cell type are given in the section entitled, "Developing a Protocol for Introducing CPA Solutions."

Similar problems are faced upon removal at post-thaw since most cells need to be washed or diluted prior to use. The frozen and thawed cells have equili-brated with the cryopreservation solution, which means that the osmolarity of the internal solution is high and the cytoplasm also contains small molecule CPAs. Cells are then transferred from a high osmolarity solution to a low osmolarity solution. Water rushes in rapidly, resulting in a rapid growth in cell volume, followed by slower removal of the small molecular weight cryopro-tective agent (Figure 3.1c). Cells are more sensitive to expansion than shrink-age, so cell losses can be more significant upon removal. There are easy, practical steps that can be used to reduce cell losses upon introduction and removal of CPAs.

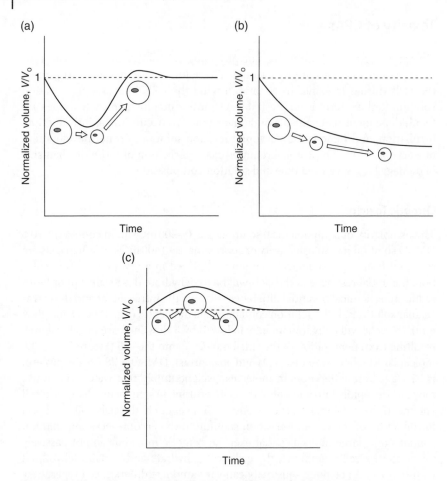

Figure 3.1 Volumetric response of cells with introduction of a (a) penetrating cryoprotective agent; (b) non-penetrating cryoprotective agent, and (c) upon removal of the cryoprotective agent after thawing Vo is the initial volume of the cells and V is the volume of the cell at time, t.

Biochemical Toxicity

CPAs, such as DMSO, are organic solvents and cells incubated in those additives will exhibit a decrease in viability with time. DMSO alters the cytoskeleton, shifts cell metabolism, and changes membrane fluidity (Fahy 1986). These effects can result in a loss of cell viability with time of exposure. Loss of viability resulting from exposure to DMSO can happen rapidly with time post-introduction. For example, it is common when cryopreserving hematopoietic stem cells (bone marrow, umbilical cord blood, and peripheral blood progenitor cells) for therapeutic applications to limit exposure to DMSO to 30 min prior to starting the freezing process in order to minimize cell losses resulting from DMSO toxicity. Biochemical toxicity can be reduced if the temperature of the cells and solution is reduced (e.g., put on ice). It is common for solutions containing DMSO to be introduced at low temperatures.

Developing a Protocol for Introducing CPA Solutions

As described above, cryoprotective solutions are not physiological solutions, so simply introducing or removing the solution can result in cell losses (before the cells are even frozen). Developing a method for introducing and removing cryoprotective solutions is an important step for a cryopreservation protocol. In general, cells can tolerate a four-fold increase in extracellular osmolarity. Therefore, cells should tolerate single step introduction into a solution less than 1200 mOsm without cell losses resulting from osmotic stress. Characterizing cell losses from biochemical toxicity is more difficult, and there are no general rules to manage cell losses.

A basic experiment can be used to characterize the response of the cells to a given introduction protocol [composition of solution, method of introduction (single step or gradual introduction), and temperature at which the introduction takes place]. The outcome of the study can then be used to understand the potential mechanism of cell losses during introduction and then develop potential strategies for mitigating the stress or cell losses. Before performing the basic experiment, it is important to select an upper threshold for the acceptable level of cell losses associated with the introduction of the solution. In most cases, it should be possible to minimize cell losses associated with the introduction of the cryopreservation solutions and keep losses to less than 10% of the cells. For cells that are sensitive to stresses of introduction, the acceptable level of cell losses may need to be higher.

The Basic Experiment

A given cryopreservation solution is introduced in a single step into a cell suspension. The sample is incubated at a given temperature and time. Samples of the cells are taken at regular time points and assayed for viability.

Case Study 1: The viability of a cell suspension was checked and high (97%) before introduction of the solution. After introduction of the solution, cell viability/cell recovery was measured at 0, 1, 2, and 4 h. The viability measured was 83, 82, 85, and 80% at the respective time points.

Interpretation and strategies to mitigate losses: Most of the cell losses occurred initially (at $t = 0$), and the viability was largely constant (within normal variations for cell viability measurements) over the 4-h time period tested. This outcome suggests that the cell losses resulted from osmotic toxicity and not from biochemical toxicity. This particular cell type cannot tolerate the step change in concentration that was tested. There are two options for reducing the cell losses: (i) adding the solution using a syringe pump or other device that slowly adds the solution to the cell suspension and therefore slowly increases the solution osmolarity or (ii) adding the solution in intermediate steps. Reducing the temperature at which the solution is added can slow down the rate of volume changes and potentially reduce cell losses as well.

Case study 2: The viability of a cell suspension was checked and high (97%) before introduction of the solution. After introduction of the solution, cell viability/cell recovery was measured at 0, 1, 2, and 4 h and was 95, 92, 85, and 80%, respectively.

Interpretation and strategies to mitigate losses: Cell losses were low initially (at $t = 0$), but losses increased steadily with time of exposure. This outcome suggests that cell losses result from biochemical toxicity. Reducing cell losses can be achieved by decreasing the time that the cells are exposed to the solution. This approach typically involves introducing the solution and then placing the cells into the freezing environment as quickly as possible. Reducing the temperature at which the cells and solution are introduced is another strategy for reducing cell losses.

Introduction of Vitrification Solutions

When determining methods for introducing vitrification solutions (or high concentration cryopreservation solutions), volumetric changes exhibited by the cells in response to transmembrane osmotic differences become important if not critical (see Figure 3.1). One goal for the introduction is to have no net cell volume change after introduction of a penetrating cryoprotective agent. When introduced to a solution formulated properly, the water will leave the cell rapidly followed by gradual introduction of the penetrating cryoprotective agent (Figure 3.1a). The initial and final cell volume should be the same.

It is important for a vitrification solution to maintain isotonicity. When CPAs are added to a carrier solution, the solution is diluted. For a 10% DMSO solution, this dilution effect is minor but for higher concentrations of CPAs, this dilution effect can be significant. A simple procedure outlined in more detail in Fahy and Wowk (2015) can be used to maintain isotonicity of the carrier solution after addition of CPAs. Briefly, a concentrated version of the carrier solution will be made and an equal weight of the concentrated carrier solution and cryoprotectants will be added and then the solution will be diluted with water to achieve the desired final volume.

The overall goal is to reduce the number of steps required for the introduction of the vitrification solution and the overall time for the process. As with conventional cryopreservation solutions, most cells can tolerate a four-fold change in tonicity of the solution without cell losses. Each step for introducing a high-concentration solution involves increasing the tonicity of the solution four-fold until the final concentration is reached.

Understanding and modeling of the osmotic response of a cell have led to the development of other approaches for adding high-concentration solutions. For example, a four-fold increase in concentration is used in a carrier solution that is one-half isotonic (Meryman 2007). This approach permits the concentration to be raised by eight-fold (rather than four-fold). Similar approaches use mathematical modeling of osmotic response of the cells to minimize shrinkage and exposure time for multistep introduction protocols (Karlsson et al. 2014). If standard introduction methods using the four-fold rule do not result in acceptable losses, it may be helpful to use mathematical modeling to optimize introduction of vitrification solutions.

Cell Concentration

Another consideration during the introduction of a cryopreservation solution is the final concentration of cells after introduction of the solution. In general, it is desirable to have the cell concentration as high as possible to reduce storage volumes and therefore storage costs. There is an upper limit to the cell concentration that can be used before there is a loss of cell viability. At very high cell concentrations, cells fuse together and lyse during thawing (Pegg et al. 1984). The water inside the cell becomes a significant fraction of the water of the sample at high cell concentrations, which can also enhance the incidence of intracellular ice formation. A cytocrit (volume of cells divided by the total volume of cells and solution) that is greater than 20% results in higher cell losses during freezing (Levin, Cravalho, and Huggins 1977). With this rule, the upper threshold of cell concentration will vary with the size of the cell. The first step is to calculate the volume of a cell, V_{cell},

$$V_{cell} = \left(\frac{4}{3}\right) \times \pi r^3$$

where r is the cell radius.

The maximum cell concentration, C, can then be calculated based on a maximum permitted cytocrit of 20% (0.2).

$$C = \frac{0.2}{V_{cell}}$$

For a cell diameter of 10–20 μm (a common range for mammalian cells), the corresponding maximum concentration is $30–50 \times 10^6$ cells/ml.

It is not uncommon for cells to be frozen at low concentrations when cell supplies are limited or the cell numbers needed are low. There is no cryobiophysical reason for low cell concentrations to increase cell losses. However, at low cell concentrations, methods of enumerating cells become less precise and volume losses associated with liquid transfer may also disproportionately influence cell recovery.

Removal of CPA Solution

Post-thaw, cells are in a high-concentration solution (i.e., 1.4 M for a 10% DMSO solution). For most applications, the cells must be transferred into a physiological solution (approximately 300 mOsm) for downstream use. Therefore, the same osmotic stresses are present at removal as were present upon introduction. Cells being removed from cryopreservation solutions will experience swelling as water enters the cell membrane in response to the difference in chemical potential between the inside and the outside of the cells. Slowly, penetrating CPAs will leave the cell and the volume will decrease (Figure 3.1c). Cells post-thaw are sensitive to volumetric excursions having been subjected to osmotic stresses during freezing and therefore can be very sensitive to removal of CPAs.

There are two common strategies for removing cryoprotective solutions with minimal cell losses. One option is to slowly dilute the cells with an isotonic solution, followed by washing the cells to remove the CPAs. Another option is to use a specially designed wash solution that is not isotonic but with a slightly higher osmolarity (approximately 600 mOSM) through the addition of large molecular weight sugars or polymers. The penetrating cryoprotective agent (i.e., DMSO) will leave the cells, as there is no DMSO in the wash solution. The higher extracellular osmolarity of the wash solution will minimize the influx of water and therefore the volume change experienced by the cell.

Safety Considerations for Cryopreservation Solutions

Cryopreservation solutions may contain components with specific safety concerns. When using proprietary cryopreservation solutions, safety recommendations should be available from the manufacturer. Specific safety measures should be developed in consultation with the safety personnel for a given organization when using cryopreservation solutions formulated in-house.

DMSO easily penetrates skin and has the potential to carry other chemicals with it. As a result, safety measures when using DMSO for cryopreservation involves limiting skin contact. Specific safety measures will vary with the volume/concentration of DMSO being used. For small volumes and limited duration of contact, rubber or nitrile gloves may be appropriate. Contact with larger volumes or for longer times typically involves the use of butyl rubber gloves. The use of safety goggles or use of a chemical fume hood for handling of DMSO is also recommended. Wearing a laboratory coat also helps prevent skin contact.

Cryopreservation Containers

The cells and cryopreservation solutions must be placed in a container that is suitable for storage at low temperatures and can maintain integrity of the sample for the duration of storage, which could be decades. Therefore, high-quality containers for cryopreservation are critical. Containers suitable only for low-temperature storage should be used and the manufacturer's specifications should be checked to ensure that the container is suitable for use at the desired storage temperature ($< 150°C$).

Three different types of containers are commonly used for cryopreservation. On a basic level, these containers are used for different volumes of samples. Straws are used for small volumes (less than 1 ml), vials are used for intermediate volumes (1–5 ml), and bags are typically used for larger volumes (5–350 ml). Other considerations that enter into the selection of a container include whether or not the device can be used for sterile closed system processing (cell therapy or cell-banking systems) or are compatible with automation.

For products that are regulated (i.e., cell therapy products), cryopreservation containers are designated as medical devices and must be manufactured in compliance with FDA Good Manufacturing Practices and International Standards Organization. As a result, companies that manufacture cryopreservation containers must register with the FDA and should submit a Device Master File or apply for 510 (k) clearance for their product.

Straws: Most commonly used for the preservation of sperm and embryos/oocytes, straws are long and thin (130–135 mm long × 1–3 mm in diameter). The high aspect ratio (length/diameter) of the sample implies that high rates of heat transfer can be experienced by the sample during freezing (500–2000°C/min). Therefore, it is common for vitrified samples to be preserved in straws. Certain straws include plugs at one end that enable efficient loading of the sample but results in a non-closed system. If there are concerns about cross-contamination or sample integrity, there are closed loading systems available for straws, and double straw systems can be used to ensure that the sample is closed.

Vials: This container is probably the most commonly used container in cryopreservation. Vials are relatively inexpensive: boxes, racks, and other containment systems are readily available for this format. Certain vials are suitable for use with automated, liquid-handling systems.

Vials consist of two components: the body where the cell suspension is located and a top (or cap), which is typically a plastic piece with threading (either internal or external) that screws onto matching threads on the top of the body. At low temperatures, the threads/seal of the vial can fail. If stored in the liquid phase of liquid nitrogen (LN_2), liquid can seep into the vial. When the vial is thawed, the LN_2 in the sample vaporizes rapidly and pressures build up inside the vial until it fails (typically in a pretty dramatic fashion). As a result, most manufacturers will specify that vials must not be stored in the liquid phase of LN_2.

Bags: Typically, bags are used to contain larger volumes/cell numbers (6–600 ml of cell suspension). Bags typically consist of a chamber for containing the cell suspension to be frozen and inlet and outlet tubing. Certain bag types may also include a pouch for labeling of the unit. It is common for cryopreservation containers to have one or more ports for introducing or removing cells, and these ports may also include different connectors such as luer locks that enable connections between the bag and other bags or equipment. For high-value cell products such as cell banks or cell therapies, processing is frequently aseptic. The inlet tubing must permit sterile connection of the cryopreservation bag with another bag or container to permit sterile transfer of the cells. Similarly, if the cells are to be manipulated post-thaw, outlet tubing must also permit sterile transfer of the cells.

Bags are typically placed into a cassette prior to freezing. The cassette consists of a flat case, most often made of aluminum or stainless steel. The high thermal conductivity of the cassette improves heat removal from the bag during freezing. The thickness of the bag is controlled by the cassette and kept consistent across the bag (typically 4–10 mm). Once again, the cassette compresses the bag to create a uniform, thin layer of cell suspension improves heat transfer from the bag and helps to ensure that the cells in the center of the bag experience a cooling rate similar to that of the cells located at the periphery of the bag. Cells experiencing different cooling rates during freezing may respond differently, and the steps that facilitate a consistent cooling rate across the sample improve consistency of the freezing protocol. Racking inside of LN_2 storage units are typically designed to accommodate cassettes and the cassette then facilitates organization of the frozen units in the repository.

Overwraps

The primary sample container (vial or bag) may be placed inside a secondary container or overwrap. The use of an overwrap can prevent contamination of the sample with infectious agents present in the LN_2. Cases of viral transmission in LN_2 have been documented (Tedder et al. 1995) and over-wraps can be used to prevent this transmission. It is common for overwraps to be made of material similar to that of the bag, as both have to tolerate freezing and low-temperature storage. Overwraps may also be used to prevent loss of product upon thawing that results from cracking or failure of the primary product bag. The use of overwraps will alter the heat transfer characteristics of the sample. Specifically, it will slow the cooling rate and the thawing rate that may be experienced by the sample (versus the same sample without an overwrap). Therefore, the use of an overwrap may require modification of a cooling or thawing protocol, in particular if it was developed without an overwrap present, to compensate for the additional resistance to heat transfer.

Labeling

The sample is only useful if it is labeled properly so that it can be identified. As with storage containers, the label must identify the sample and be stable for the duration of storage, which can be weeks, months, and even decades. A wide variety of labeling options are available for cryopreservation containers. Labeling of samples to be stored at low temperatures is challenging. Handwriting on the outside of a cryopreservation container can be obscured by frost and then accidentally removed when the frost is removed. Adhesives tend to perform poorly at low temperature, and externally applied labels can fall off during storage.

Handwritten labels: A very common practice in laboratories is to write the sample identification on the surface of the bag or cryovial. If handwritten labels are to be used, the ink should be solvent resistant and stable at low temperatures.

Pre-labeled vials: A variety of manufacturers have vials that are preprinted with a label or bar code. This capability eliminates the need to affix a label on the surface.

Printed labels: A variety of suppliers produce preprinted labels that can be affixed to the outer surface of a container. Printed labels may include written identifiers as well as bar codes or other machine-read methods of identifying a sample.

Radio-frequency identification (RFID): This sensor uses electromagnetic fields to identify samples. RFID can be made to both identify the sample and monitor temperature of the sample.

For products that are regulated, the cryopreservation container must include (i) a tamper-evident product identifier; (ii) a lot identifier; and (iii) a manufacturer identifier. This labeling requirement enables traceability of the product and the container.

Sample Annotation

As with the pre-freeze processing, the processing record for the sample should include the following:

- Description of the composition of the cryopreservation solution
- The method by which the solution was introduced, including the following:
- Time of incubation (or limits on time exposure)
- Temperature of introduction (for both cells and solution)
- Method of introduction for the solution

Scientific Principles

- Cryopreservation solutions are formulated to alter the behavior of water and preserve critical structures in the cell during freezing.
- These solutions are not isotonic and may cause cell damage resulting from osmotic and biochemical toxicities.

Putting Principles into Practice

- Cryopreservation solutions commonly contain a carrier solution, CPAs, and sometimes a protein.
- The highest quality components should be used in the solution formulation, as cells in the solution cannot be purified and residual contaminants have been shown to influence post-thaw recovery and its reproducibility.
- Use of expired or incorrectly formulated cryopreservation solutions is a recognized source of variability in preservation protocols. Use of simple quality control measures can reduce the potential for poor outcome.
- Protocols for introducing and removing cryoprotective/vitrification solutions should be developed in order to minimize cell losses.
- Use of containers that are appropriate for low-temperature storage for up to decades is critical.
- Downstream use of the sample requires proper labeling of the sample. Labels must function at low temperatures and be useful for the duration of storage.

References

Anchordoguy, T. J., A. S. Rudolph, J. F. Carpenter, and J. H. Crowe. 1987. "Modes of interaction of cryoprotectants with membrane phospholipids during freezing." *Cryobiology* 24 (4):324–331.
Angell, C. A. 1995. "Formation of glasses from liquids and biopolymers." *Science* 267 (5206):1924–1935.

Arakawa, T., and S. N. Timasheff. 1985. "The stabilization of proteins by osmolytes." *Biophys J* 47 (3):411–414.

Bhatnagar, B. S., R. H. Bogner, and M. J. Pikal. 2007. "Protein stability during freezing: separation of stresses and mechanisms of protein stabilization." *Pharm Dev Technol* 12 (5):505–523.

Cocks, F. H., and W. E. Brower. 1974. "Phase diagram relationship in cryobiology." *Cryobiology* 11:340–358.

Fahy, G. M. 1986. "The relevance of cryoprotectant "toxicity" to cryobiology." *Cryobiology* 23 (1):1–13.

Fahy, G. M. 2010. "Cryoprotectant toxicity neutralization." *Cryobiology* 60 (3 Suppl):S45–S53.

Fahy, G. M., and B. Wowk. 2015. "Principles of cryopreservation by vitrification." *Methods Mol Biol* 1257:21–82.

Freedman, L. P., M. C. Gibson, S. P. Ethier, H. R. Soule, R. M. Neve, and Y. A. Reid. 2015. "Reproducibility: changing the policies and culture of cell line authentication." *Nat Methods* 12 (6):493–497.

Karlsson, J. O., E. A. Szurek, A. Z. Higgins, S. R. Lee, and A. Eroglu. 2014. "Optimization of cryoprotectant loading into murine and human oocytes." *Cryobiology* 68 (1):18–28.

Levin, R. L., E. G. Cravalho, and C. E. Huggins. 1977. "Water transport in a cluster of closely packed erythrocytes at subzero temperatures." *Cryobiology* 14 (5):549–558.

Lovelock, J. E., and M. W. Bishop. 1959. "Prevention of freezing damage to living cells by dimethyl sulphoxide." *Nature* 183 (4672):1394–1395.

McGrath, J. J., K. R. Diller, American Society of Mechanical Engineers. Winter Meeting, American Society of Mechanical Engineers. Bioengineering Division, and American Society of Mechanical Engineers. Heat Transfer Division. 1988. Low temperature biotechnology: emerging applications and engineering contributions, presented at the winter annual meeting of the American Society of Mechanical Engineers, Chicago, IL, November 27–December 2, 1988, BED; vol. 10. New York: The American Society of Mechanical Engineers.

Meryman, H. T. 2007. "Cryopreservation of living cells: principles and practice." *Transfusion* 47 (5):935–945.

Pegg, D. E., M. P. Diaper, H. L. Skaer, and C. J. Hunt. 1984. "The effect of cooling rate and warming rate on the packing effect in human erythrocytes frozen and thawed in the presence of 2M glycerol." *Cryobiology* 21 (5):491–502.

Polge, C., A. U. Smith, and A. Parkes. 1949. "Revival of spermatozoa after vitrification and dehydration at low temperatures." *Nature* 164:666.

Pollock, K., G. Yu, R. Moller-Trane, M. Koran, P. I. Dosa, D. H. McKenna, and A. Hubel. 2016. "Combinations of osmolytes, including monosaccharides, disaccharides, and sugar alcohols act in concert during cryopreservation to improve mesenchymal stromal cell survival." *Tissue Eng Part C Methods* 22 (11):999–1008.

Pollock, K., R. M. Samsonraj, A. Dudakovic, R. Thaler, A. Stumbras, D. H. McKenna, P. I. Dosa, A. J. van Wijnen, and A. Hubel. 2017. "Improved post-thaw function and epigenetic changes in mesenchymal stromal cells cryopreserved using multicomponent osmolyte solutions." *Stem Cells Dev.* 26(11): 828–842.

Sputtek, A. 1991. "Cryopreservation of red blood cells, platelets, lymphocytes, and stem cells." In Clinical Applications of Cryobiology edited by B. J. Fuller and B. W. W. Grout, 95–147. Boca Raton, FL: CRC Press.

Tedder, R. S., M. A. Zuckerman, A. H. Goldstone, A. E. Hawkins, A. Fielding, E. M. Briggs, D. Irwin, S. Blair, A. M. Gorman, K. G. Patterson, D. C. Linch, J. Heptonstall, and N. S. Brink. 1995. "Hepatitis B transmission from contaminated cryopreservation tank." *Lancet* 346 (8968):137–140.

Vanderkooi, J. M., J. L. Dashnau, and B. Zelent. 2005. "Temperature excursion infrared (TEIR) spectroscopy used to study hydrogen bonding between water and biomolecules." *Biochim Biophys Acta* 1749 (2):214–233.

4

Freezing Protocols

Importance of Cooling Rate

As described in Chapter 1, freezing acts to preserve cells by changing the mobility of water and the activity of degradative molecules. Thus, cells must be cooled from physiological temperatures (approximately 37°C for mammalian cells) to cryogenic temperatures for storage. The rate at which cells are cooled (°C/min) has a strong influence on their survival. It has been known since the 1960s (Leibo and Mazur 1971, Mazur 2004) that there is a strong variation in post-thaw survival with cooling rate (B) during freezing and that relationship can vary with cell type and composition of the freezing medium (Figure 4.1).

The variation in survival with B most often takes the shape of an inverted "U." Both "low" and "high" cooling rates are associated with poor survival and there is a narrow range of cooling rates associated with maximal survival. The range of cooling rates associated with maximal survival will vary from cell type to cell type (Figure 4.1a). For example, RBCs in low concentrations of CPAs exhibit high survival for cooling rates approximately 2000°C/min, whereas hematopoietic stem cells exhibit high survival at 1°C/min (Mazur et al. 1970).

Survival of a specific cell type will also vary with composition of the freezing medium (Figure 4.1b). In general, the higher the concentration of CPAs, the slower the cooling rate associated with optimal survival.

Cooling samples is performed using two basic methods: controlled-rate freezing and passive freezing. The processes of constructing controlled-rate and passive-freezing protocols are given in the following. In general, when high levels of survival and reproducible post-thaw recovery are desired, controlled-rate freezing protocols are used. For resource-constrained environments or where overall recovery and reproducibility are not critical, passive-freezing protocols may be appropriate.

Preservation of Cells: A Practical Manual, First Edition. Allison Hubel.
© 2018 John Wiley & Sons, Inc. Published 2018 by John Wiley & Sons, Inc.

(a)

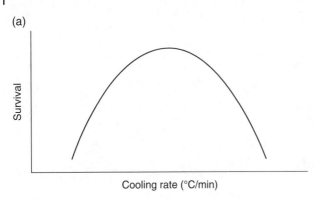

(b)

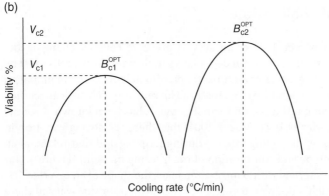

Figure 4.1 (a) Variation in survival with cooling rate for a given cell type. (b) Variation in survival with cooling rate for a specific cell type but varying composition of cryopreservation medium. V_{c1} and V_{c2} are the post-thaw viabilities associated with compositions 1 and 2, respectively. B_{c1}^{OPT} and B_{c2}^{OPT} are the optimum cooling rates associated with compositions 1 and 2, respectively.

Controlled-rate Freezing

There are several different types of controlled-rate freezers. Designs vary principally through the manner by which the sample is cooled, the number of samples that can be frozen together, and their geometry (bags, vials, straws). The scope of this section will be limited to devices that are currently commercially available and in fairly widespread use.

One approach to cooling a sample involves the use of the Stirling refrigeration cycle to remove heat from a sample (Lopez, Cipri, and Naso 2012). Briefly, a Stirling refrigeration engine behaves as a heat pump. Heat is in effect pumped out of this compartment to be cooled, through the working gas of the cryocooler, and into the compression space. The compression space will be above ambient temperature, and so heat will flow out into the environment. This approach does not require use of liquid nitrogen (LN_2), but a source of electricity to achieve temperatures down to about −100°C. This approach works well for environments in which a supply of LN_2 is not available or desirable. There are limitations in terms of the cooling rate that can be achieved (0.1–15°C/min), and Stirling engines are associated with vibrations, which can cause damage to the cells during freezing (Lopez, Cipri, and Naso 2012).

Another approach uses LN_2 to provide open-loop cooling of the sample and a heater provides compensatory heating for the sample. The temperature of the sample decreases as the power to the heater decreases as specified by a control unit. As with the example described above, there are limitations to the cooling rate that can be achieved (0.01 to roughly 10°C/min).

A third method of controlled cooling involves taking chilled nitrogen gas and circulating it in a chamber containing the sample to be cooled. A control system detects the temperature of the chamber and increases or decreases the flow rate of the coolant based on the program that has been entered into the system. This approach can achieve a wider range of cooling rates (0.1–50°C/min) and volumes (up to 48 l).

It is likely that new controlled-rate freezing devices will be commercialized in the future. There is a movement toward standardization and development of a manufacturing paradigm for cell therapies and regenerative medicine products. This movement will result in the development of automated technologies that are integrated into the overall workflow. The principles for the construction of a controlled-rate protocol given in the following can be used independent of the method of cooling.

Controlled Cooling-rate Protocols

Controlled-rate freezers cool the sample by reducing the temperature of the chamber containing the sample to be frozen in a controlled fashion. Development of a controlled-rate protocol requires specifications of the various cooling or holding steps for the protocol shown schematically in Figure 4.2.

Segment 1: Initial Hold Period
Designing Segment 1 involves selecting a temperature and duration for the initial hold period. It is common for the freezing chamber of a controlled-rate freezer to be prechilled to a temperature slightly higher than the freezing temperature of the cells (0–4°C). The prechilling of the chamber has a couple of major functions: (i) to keep cells cool and thereby reduce the potential for cell losses resulting from exposure to the cryoprotectants and (ii) cooling the sample from room temperature (approximately 20°C) using a controlled-rate freezer does not provide any net benefit (in terms of viability) and can increase the overall time for the protocol.

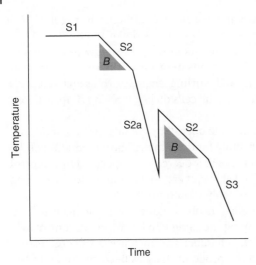

Figure 4.2 Temperature as a function of time for the freezing chamber during a controlled-rate protocol. The programmed elements of a controlled-rate protocol include Segment 1 (S1), initial hold; Segment 2 (S2), controlled cooling rate; Segment 2a (S2a), nucleation step (optional); and Segment 3 (S3), higher cooling rate (optional) to final temperature where sample is removed and placed in storage. "B" is the cooling rate for the sample, which is typically specified in the protocol.

Segment 1 should be sufficiently long to permit equilibration of the sample with the chamber. If the cells being preserved are processed at room temperature, the hold period will be greater to permit the cells to cool. If a large volume of cells is frozen (e.g., in a single bag or even in multiple small vials), the hold period will increase. A hold period of 10–15 min is not uncommon.

The temperature as a function of time for the sample measured with the sample probe or an external temperature measurement device inserted into the freezing chamber for the purposes of measuring the temperature of the sample will be important in validating Segment 1. If Segment 1 is properly designed, the temperature of the sample will decrease after being placed in the chamber and equilibrate with the chamber (Figure 4.3b, dotted line). At the beginning of Segment 2, the sample and chamber temperatures will track fairly closely together until the sample undergoes nucleation (the formation of a self-assembled state of water resembling the structure of ice). If Segment 1 is too short, the initial temperature of the sample will not equilibrate with the chamber (Figure 4.3b, dashed line). Then, as Segment 2 begins, the sample temperature will lag significantly behind the chamber temperature (e.g., it will be higher) and, more importantly, the temperature will vary from sample to sample.

Segment 2: Cooling

The second segment for a controlled-rate freezing protocol starts at the holding temperature used in Segment 1, with a decrease in the chamber temperature as a function of time during this segment. When a protocol specifies a cooling rate (e.g., 1°C/min), it is the cooling rate in this segment that is being specified. Damage (or not) to the cells is most often observed over this range of

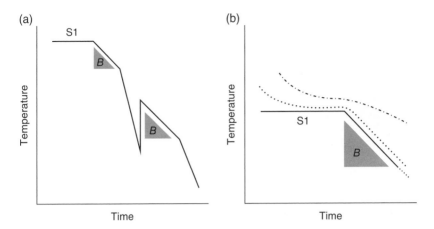

Figure 4.3 (a) Segment 1 (S1) as part of the overall protocol. (b) Expanded view of Segment 1. The chamber temperature is given as a solid line. If the time for equilibration is sufficient and the sample equilibrates with the surrounding chamber, then it will track with the initial portion of Segment 2 (S2) (dotted line). If the time of equilibration is insufficient, the sample temperature will not track with the chamber temperature (dotted and dashed lines).

high subzero temperatures and, as a result, the cooling rate in this region is the most critical to control. Segment 2 continues until the sample is at a considerably lower temperature that is determined based on the temperature for onset of Segment 3.

As the sample cools, the cells and surrounding aqueous solution remains largely unchanged until ice is formed. The nucleation of ice and subsequent crystal growth removes water from the solution. The biological cells and solute is rejected from the growing ice phase and cells are sequestered into the gaps between adjacent ice crystals and surrounded by high-concentration liquid (Figure 4.4). The growth of the ice phase changes both the chemical and the mechanical environment of the cells and, as a result, the temperature at which ice nucleates and grows is important to the survival of cells during freezing.

Pure water during cooling *does not* freeze at 0°C (however, it melts at 0°C). It can freeze over a range of temperatures ranging from 0 to as low as −40°C (Hobbs 1974). For most practical situations (large volumes, presence of solutes), a cell suspension will freeze in the range of temperatures from −5 to −15°C. The nucleation process is stochastic, implying a distribution of possible outcomes is normal and expected. If 100 vials each containing 1 ml of a cell suspension are placed into a controlled-rate freezer and the samples are instrumented with thermocouples to measure the temperature during freezing, and the vials are cooled at a constant cooling rate to a low temperature (e.g., −40°C), a small number of vials would nucleate at high subzero temperatures (−3 to −5°C), the majority of vials would nucleate at intermediate

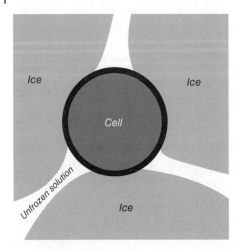

Figure 4.4 The cell remains in the unfrozen liquid between adjacent ice crystals. As the temperature decreases, the size of the gap between adjacent ice crystals decreases and the concentration of the unfrozen solution increases.

temperatures (−5 to −9°C), and a small number of vials would nucleate at low temperatures (−9 to −15°C). *The temperature at which nucleation takes place will vary from sample to sample and from freezing run to freezing run. This variation reflects the physics of the nucleation process (Toner, Cravalho, and Karel 1990).*

As described above, distinct changes in the chemical and mechanical environment of the cells takes place after the sample is nucleated. Cells respond to the changes in environment (increased concentration of solution) by expressing water to reduce the difference in chemical potential between the inside of the cell and the outside of the cell. The ability of water to leave the cell is strongly influenced by temperature. Specifically, the permeability of the cell membrane to water decreases with decreasing temperature (Mazur 1963). The consequence of this relationship implies that the lower the temperature for nucleation of the sample, the greater the amount of water in the cell during freezing, and the greater the potential for ice formation inside the cell. Decreasing the temperature at which ice forms in the extracellular solution increases the fraction of cells that are dead for a given cooling rate (Toner 1993, Toner, Cravalho, and Karel 1990). This relationship has been demonstrated for a variety of cell types. Therefore, it is common for controlled-rate freezing protocols to include a step to influence or control nucleation of the sample. This relationship implies that it is important to minimize the *undercooling* of the cells. Undercooling of the cells is the temperature difference, ΔT, between the melting temperature of the sample, T_m, and the temperature at which ice forms in the extracellular solution, T_{nuc}. Decreasing the temperature at which ice forms in the extracellular solution increases the undercooling of the cells and the potential for cell damage.

Segment 2a: Seeding/nucleation of the sample. As described earlier, the nucleation of the sample brings about significant changes in the chemical and mechanical environment of the cell. There are three basic methods for seeding or nucleating the sample: (a) uncontrolled nucleation, (b) manual seeding, and (c) automatic seeding (Figure 4.5).

Uncontrolled Nucleation

Samples in which Segment 2 just contains a constant cooling rate without manual seeding or automatic seeding steps are considered uncontrolled nucleation (Figure 4.5a). Samples will nucleate spontaneously. Because of the

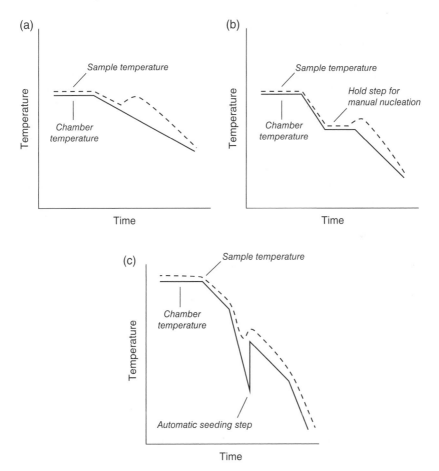

Figure 4.5 Expanded views of Segments 1 and 2 for the controlled-rate protocol demonstrating the three different methods of nucleation: (a) uncontrolled nucleation, (b) manual nucleation, and (c) automatic nucleation. The solid line represents the temperature of the cooling chamber. The dashed line represents the temperature of the sample.

stochastic nature of the nucleation process, T_{nuc} will vary significantly from sample to sample. Variation in T_{nuc} between −5 and −15°C for cells frozen in bags could be expected. Certain cell types can tolerate significant undercooling (e.g., lymphocytes), while other cell types respond very poorly to uncontrolled nucleation (e.g., hepatocytes). A constant cooling rate for the entirety of Segment 2 is not commonly seen amongst controlled freezing protocols and does not use the power of the controlled-rate freezer. Uncontrolled nucleation is typically seen in passive-freezing protocols described in the following.

Manual Nucleation

During manual nucleation, the sample is cooled to a given subzero nucleation temperature during Segment 2 and then held at that fairly high subzero temperature. The freezing chamber is opened and the sample(s) are nucleated (Figure 4.5b). One method of manually nucleating the sample involves touching the sample with a metal object that has been cooled in LN_2 to induce a localized cooling that will induce nucleation. An alternative method of localized cooling involves spraying a sample with a narrow stream of LN_2. After nucleation has occurred, the chamber is closed, the hold step in the controlled cooling protocol is removed, and the cooling process continues until the final temperature for Segment 2 has been reached. Manual seeding is performed most often on samples that are very sensitive to undercooling such as embryos.

Automatic Nucleation

One of the most common methods used to seed or nucleate the sample during freezing involves rapid cooling of the sample to a low temperature followed by a rapid warming (Figure 4.5c). This series of segments is often described as a seeding step or automatic nucleation. As with the manual freezing, the intention is to control the temperature at which ice forms in the extracellular space. It is noteworthy that at the end of the warming step, the final temperature of the chamber is programmed to be the same as if the protocol had not contained the step.

As described earlier, the manual seeding step is designed to induce a localized cooling that will induce nucleation. The intent of the automatic seeding step is to induce that same effect (localized cooling) by rapidly cooling and rapidly warming the sample.

Determining when the sample has seeded/nucleated and whether there is ice in the extracellular space is fairly straightforward. Nucleation and crystal growth result in the release of the latent heat of fusion as the water molecules in the sample transition from a liquid to a solid. The release of the latent heat of fusion results in a temperature rise that can be easily detected using probes in or adjacent to the sample being frozen (Figure 4.6). The temperature rise and

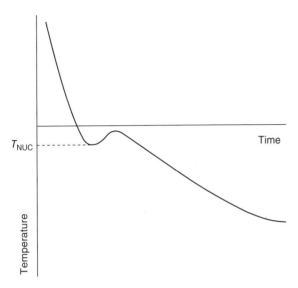

Figure 4.6 Identifying the temperature at which ice forms in the extracellular solution. The sample cools to a temperature below the melting temperature. The release of the latent heat of fusion results in an increase in the temperature. The point at which the temperature increase starts can be considered the T_{nuc}.

its duration will vary based on the volume being frozen and the cooling rate; the greater the volume being frozen, the longer the duration for the release of the latent heat.

The temperature at which ice forms (and not the time) is important.

Verifying Segment 2 (Including S2a)

If the cooling rate selected for Segment 2 is improper (too high or too low), the cells will die. If the automatic seeding segment is improper (the rapid cooling followed by rapid warming of the sample), the cells will die. This fact implies that terminating the freezing protocol, thawing the cells, and determining post-thaw viability can help determine if that particular segment is optimized. The freezing process can be stopped and the sample can be thawed during Segment 2 before the automatic seeding step or after the sample has cooled and warmed to determine whether the seeding step influenced viability.

"Delayed" Latent Heat

As indicated previously, the nucleation process is stochastic and can take place over a range of temperatures. The addition of a nucleating step is intended to narrow the range of temperatures over which nucleation takes place. The use of automatic seeding steps does not ensure that the sample will actually seed at a given time or temperature during the freezing run. In many freezing protocols, it is expected that the release of the latent heat occurs at a specific time during the protocol. If the nucleation and release of the latent heat is observed later than expected (i.e., delayed latent heat), it is considered a deviation in the protocol, and post-thaw viability testing of the sample may be required to determine whether or not the delay in the release of the latent heat is associated with reduced post-thaw recovery.

There is much confusion regarding this aspect of a freezing curve. Some common errors or confusions will be discussed here.

It is important to keep in mind what is being measured: the temperature of the sample or vial containing the cryopreservation solution (i.e., a dummy sample) that is instrumented with a temperature measurement device. If multiple samples are frozen, the only sample whose nucleation characteristics are known is the one that is instrumented. One sample in the chamber experiencing delayed latent heat (e.g., delayed nucleation or higher undercooling) does not imply that all samples experienced the same phenomena.

It is the temperature for nucleation that is critical and not the time. Protocols should specify the lowest acceptable temperature for nucleation and not the time. The time for onset of nucleation may vary with the load placed in the freezer (e.g., a single vial versus 100 vials). Deviations from the protocol should be based on the temperature for the onset of nucleation (T_{nuc}) and not on the time for onset of nucleation. A common error is to specify a time (or time window for nucleation) rather than a range of temperatures over which nucleation should be experienced.

Segment 3

The third segment of the freezing protocol involves cooling the sample from a low temperature (about $-50°C$) to the final temperature of the controlled-rate protocol (about $-100°C$) using a higher cooling rate (Figure 4.2). The response of the cell to the freezing environment diminishes with decreasing temperature. There is a point at which the sample can be cooled more rapidly (thereby decreasing the time for the protocol and cooling costs) to the final temperature of the controlled-rate freezing protocol. The final temperature for the controlled-rate protocol is selected to prevent significant warming of the sample when it is transferred from the controlled-rate freezer to a storage unit.

Verifying Segment 3

The two critical issues for formulation of Segment 3 are the temperature at which the segment begins and the cooling rate. Therefore, verifying this segment can involve cooling the sample to a given final temperature for Segment 2 (e.g., $-50°C$) and removing one sample, thawing that sample, and determining viability. A remaining sample is cooled at a higher cooling rate to the final temperature (approximately $100°C$) and then thawed. If viabilities are comparable between the two samples, the segment is appropriate. If the sample thawed at the end of Segment 3 has a lower viability than that thawed at the end of Segment 2, the temperature for the start of Segment 3 and/or the cooling rate for Segment 3 should be reduced. The desired outcome is for the viability at the end of Segment 3 to be comparable to that determined at the end of Segment 2.

Other Types of Controlled-rate Protocols

The description given previously is intended for constant-rate protocols. Other types of protocols may include nonlinear cooling rate protocols. These protocols are typically created by linking biophysical modeling of a given cell type with modeling of mechanisms of damage. The resulting protocol is intended to minimize the time for cryopreservation and the formation of ice inside the cell. This type of protocol is not currently widely used.

Passive Freezing

Passive freezing (i.e., dump freezing) involves taking a sample to be cryopreserved and placing it in a low-temperature environment (typically a mechanical freezer at −80°C). Unlike a controlled-rate freezer, the temperature of the freezing chamber is low (e.g., −80°C) and does not change with time. The sample cools after being placed in the chamber. The temperature difference between the sample and the chamber decreases with time and therefore the cooling rate experienced by a sample will decrease with time. The following three-step description will help the reader develop a passive-freezing protocol. There are commercially available passive-freezing devices. Their use and design are described at the end of this section.

Step 1: Preparing the passive-freezing unit. A key to a successful passive-freezing protocol is making the process as reproducible as possible. A passive-freezing protocol will be more reproducible if the heat transfer during freezing is reproducible. Therefore, the mechanical freezer should be empty and recently defrosted. The samples stored in the unit and ice buildup on the walls and shelves will alter the heat transfer and therefore the reproducibility of the freezing process.

Step 2: Determining the cooling rate during freezing. A temperature sensor and data logger will be needed for this step.

A bag/vial containing the solution to be frozen should be instrumented with a thermocouple. Inserting the temperature measurement device into the bag or drilling a hole in the cap of the vial and gluing the thermocouple into a position in the middle of the solution to be frozen give the most accurate temperature for the sample. It is also acceptable to tape or adhere the temperature measurement device to the outside of the bag or vial. When taped to the outside of the bag/vial, the cooling rate measured will be less than that actually experienced by the cells.

The instrumented sample will then be placed in the mechanical freezer and the temperature versus time during freezing should be recorded. It will be important that the bag/vial contains the same composition of freezing solution that will be used for preserving the cells. Different solution compositions will have different freezing behaviors. For example, a solution containing 10%

dimethylsulfoxide will freeze differently than an isotonic saline solution. The bag does not need to contain cells.

The cooling rate can be calculated based on the temperature as a function of time measured. There are two basic cooling rates that can be calculated: average and local/instantaneous. In both situations, the cooling rate is defined as

$$B = \frac{T_{\text{final}} - T_{\text{initial}}}{t_{\text{final}} - t_{\text{initial}}}$$

where T is the temperature and t is the time. The average cooling rate should be calculated over the temperature range for the freezing process. For passive freezing in a mechanical freezer at $-80°C$, an estimate for the average cooling rate can be calculated assuming

$$B = \frac{0°C - (-80°C)}{t(0°C) - t(-80°C)}$$

where $t(0°C)$ is the time at which the sample is at $0°C$ and $t(-80°C)$ is the time at which the sample is at $-80°C$. The use of an average cooling rate represents a risk. If a local or instantaneous cooling rate is too high, the cells may be damaged. Local or instantaneous cooling rates can also be calculated to characterize the range of cooling rates actually produced during a passive-freezing protocol. Calculating the cooling rate becomes important only after ice has started forming in the extracellular space ($T > T_{\text{nuc}}$); for most samples this will be between -5 and $-9°C$. The maximum average cooling rate for samples in a $-80°C$ mechanical freezer is between 5 and $7°C/\text{min}$. Local cooling rates can vary considerable ($1-10°C/\text{min}$). Freezing samples in a $-150°C$ freezer results in higher average cooling rate as the temperature difference is greater.

In a successful passive-freezing protocol, the optimum cooling rate for a given cell type and solution composition should be in the range that can be achieved for the equipment being used. If the cooling rate needed for acceptable post-thaw viability is *lower* than that measured for your system, the sample can be wrapped in insulation (e.g., Styrofoam) to reduce the cooling rate. Step 2 can be repeated to validate that the cooling rate has decreased to the desired level with the addition of the insulation. The thickness and composition of the insulation should be noted so that the same configuration can be repeated for freezing of subsequent samples.

If the cooling rate needed for acceptable post-thaw viability is *higher* than that measured for your system, it may not be possible to use passive freezing for your sample (assuming that a colder mechanical freezer is not available).

Step 3: Freezing cell suspensions using the desired cooling rate. As with Step 2, the samples should be instrumented with a thermocouple or similar temperature-sensing device and recorded during the freezing process. The average cooling rate, undercooling, and post-thaw viability of the sample should be determined. Typically, post-thaw recovery of cells in passive-freezing devices is less than that seen in controlled-rate freezers. Since nucleation in passive-freezing protocols is uncontrolled, it will be important to repeat Step 3 several times and quantify the influence of this factor on post-thaw recovery for the cell type of interest. If the variation is too large, the passive-freezing process may not be appropriate for intended use.

There are commercially available passive-freezing devices that are designed to produce an average cooling rate (most commonly 1°C/min) for a sample placed in the device. This outcome is accomplished through two different methods: (i) an alcohol bath or (ii) foam insulation of a given thickness and geometry. Passive-freezing devices that use alcohol require routine maintenance. Specifically, the alcohol absorbs water from the atmosphere and it must be changed regularly (see manufacturer's recommendations). There are no such requirements for foam-based passive-freezing devices.

Important caveat: The following guidelines for developing a passive or controlled-rate protocol is limited to a given cell type and a given composition of cryopreservation solution. Significant changes in composition of the medium may change the cooling rate over which optimum post-thaw viability is observed and therefore require alterations in the freezing protocol for the new composition. In addition, there is no single freezing protocol appropriate for all cell types. As described earlier, different cell types may require different cooling rates for optimal post-thaw recovery. In addition, one cell type may be more sensitive to undercooling than another, requiring differences in nucleation methods.

Transfer to Storage

After completion of the freezing protocol, the sample should be transferred from the freezing unit to a storage facility. Samples removed from controlled-rate freezers or passive-freezing devices warm rapidly when removed from the unit. The sample may look frozen, but the sample is warming 10–100°C/min during transfer, which can result in partial melting of the sample or recrystallization in the extracellular and intracellular spaces. The end result of this transient warming event is loss of viability of the cells and should be avoided.

There are a variety of commercially available sample carriers that can be used to transfer a sample from a freezing device (controlled rate or passive) to a storage unit. The temperature of the sample carrier should be as close as possible to either the temperature of the freezing device or storage. For example, if removing a sample from a controlled-rate freezer at −100°C, the temperature of the sample carrier should be less than −100°C. There is no point in using dry ice which is at a temperature of −80°C, for example, since this would result in rapid warming of the sample. Certain commercially available carriers will record the temperature as a function of time during the transfer process and this data can be incorporated into the production record of the sample.

Vitrification

Vitrification protocols typically do not involve programming a freezing chamber. Most often, they operate in a manner similar to passive freezing. Namely, the sample is placed in a very cold environment (typically LN_2) and cooled in an uncontrolled fashion to temperatures below the glass transition temperature for the solution.

The most common method of vitrifying a sample involves plunging a sample into LN_2. Straws made of quartz or small-diameter polymer straws are used to achieve high-cooling rates required for vitrification. A similar approach has been to use what is described as a cryoloop. The sample is placed in a small metal loop that is vitrified by immersion in LN_2. A third method involves placement of a sample to be vitrified in a minimal volume of solution on top of a metal plate that is then immersed in LN_2. Another method involves placing a portion of the container into LN_2 and then loading the sample into the chilled container using a pipette to vitrify. As with the other systems described earlier, this method is an open system, and there is contact between the sample and the LN_2 (Asghar et al. 2014).

One approach to improve heat transfer during vitrification involves reducing the pressure at which the process takes place, resulting in the formation of slush (a mixture of solid and LN_2). A sample in this mixture does not produce boiling, and heat transfer is enhanced. Cooling rates of 130000°C/min have been achieved (Lopez, Cipri, and Naso 2012).

In general, the methods described above are limited to small volumes and typically involve the use of high concentrations of cryoprotective agents (it is the combination of cooling rate and concentration that results in vitrification). Vitrification techniques are most commonly used with embryos or oocytes since the volume of the sample is small enough and the micromanipulation skills needed for introduction and removal of the vitrification solutions are common in the laboratory.

Independent Temperature Measurement

Since the variation in temperature as a function of time during cooling plays an important role in cell survival, measuring the temperature and recording the variation in temperature with time can be critical in developing or validating either a controlled-rate protocol or a passive-freezing protocol.

Measurement of temperature during freezing can be performed using a variety of methods. Thermocouples are often used as temperature sensors as they are rugged and inexpensive. A type "T: thermocouple (copper and Constantan) is used most often over the temperature range during freezing (0 to $-196°C$). Resistance temperature devices can also be used to measure temperature over this range. A data logger facilitates recording temperature as a function of time and thereby permits calculation of the cooling rate (change in temperature/ time). These devices are neither expensive nor difficult to operate and can be used to monitor temperature during freezing and storage.

Most controlled-rate freezers have both temperature measurement probes and a data logger for recording the temperature during freezing. There are typically two different temperature probes: the chamber probe and the sample probe. The chamber probe measures the temperature in one location of the chamber and that temperature is what is programmed into the device. The sample probe can be placed next to the sample (e.g., next to a bag in a freezing cassette) being frozen. This results in errors, as there is a temperature difference between the outside of the bag (where the probe is) and the inside of the bag (where the cells are). However, this method is commonly used.

An alternative is to use a control sample to monitor temperature during freezing. The temperature probe is placed into the cell suspension either through a port in a bag or drilling a hole in a vial and inserting the probe through the hole. In these situations, the sterility of the sample is not maintained, so the sample should not contain a cell suspension but a volume of freezing solution that mimics that used for the protocol. Freezing a control sample may not reflect the freezing behavior of all samples in the chamber, especially, when there are many samples frozen together. The cryopreservation solution in the control sample should be changed after every freezing run. The freezing process partitions water and the solutes in the cryopreservation solution. Thus, the freezing properties of a cryopreservation solution will change with subsequent freeze–thaw cycles.

The temperature of the chamber and the sample temperature will be used for development of a controlled-rate protocol (see the following). This data should be stored as part of the production record for that cell product. For a passive-freezing protocol, the temperature probes and data logger will be critical in developing a reproducible passive-freezing protocol. As with a controlled-rate protocol, the temperature as a function of time for passive freezing should be recorded and stored as part of the production record for the product.

Scientific Principles

- The rate of temperature drop for a range of temperatures from slightly above freezing to about $-60°C$ is the cooling rate. The cooling rate is an important factor in the post-thaw viability of cells.

- Changes in composition can also alter the cooling rate at which optimum post-thaw recovery is observed.
- The temperature at which ice forms (T_{nuc}) in the extracellular solution influences cell response. Lower values of T_{nuc} for a given cooling rate results in lower post-thaw recoveries.

Putting Principles into Practice

- Controlled-rate freezers provide more control of the freezing process and a controlled cooling-rate protocol consists of a series of steps that can be designed using rational methods in order to achieve favorable post-thaw recovery.
- Passive freezing can be used with cell types that are not sensitive to undercooling and for a limited range of optimum cooling rates.
- Vitrification of a sample typically involves plunging a sample into LN_2. Small volume systems are most often vitrified due to heat transfer limitations.
- Having an independent method of measuring and monitoring temperature (most often thermocouples and a data logger) is extremely helpful when developing or verifying a freezing process, regardless of the process used.
- Commercially available sample carriers can be used to transfer the product from the freezing device to storage without resulting in warming of the sample.

References

Asghar, W., R. El Assal, H. Shafiee, R. M. Anchan, and U. Demirci. 2014. "Preserving human cells for regenerative, reproductive, and transfusion medicine." *Biotechnol J* 9 (7):895–903.

Hobbs, P. V. 1974. "Nucleation of ice." In *Ice Physics*, edited by P. V. Hobb, 461–523. Oxford, England: Oxford University Press.

Leibo, S. P., and P. Mazur. 1971. "The role of cooling rates in low-temperature preservation." *Cryobiology* 8 (5):447–452.

Lopez, E., K. Cipri, and V. Naso. 2012. "Technologies for cryopreservation: overview and innovation." In *Current Frontiers in Cryobiology*, edited by I. Katkov, 527–546. Rijeka, Croatia: Intech.

Mazur, P. 1963. "Kinetics of water loss from cells at subzero temperature and the likelihood of intracellular freezing." *J Gen Physiol* 47:347–369.

Mazur, P. 2004. "Principles of cryobiology." In *Life in the Frozen State*, edited by B. J. Fuller, N. Lane, and E. Benson, 3–66. Boca Raton, FL: CRC Press.

Mazur, P., S. P. Leibo, J. Farrant, E. H. Y. Chu, M. G. Hanna, and L. H. Smith. 1970. "Interactions of cooling rate, warming rate and protective additives on the survival of frozen mammalian cells." In *The Frozen Cell*, edited by G. E. W. Wolstenholme and M. O'Connor, 69–88. London, UK: J&A Churchill.

Toner, M. 1993. "Nucleation of ice crystals inside biological cells." In *Advances in Low Temperature Biology*, edited by P. Steponkus, 1–51. London, UK: JAI Press.

Toner, M., E. G. Cravalho, and M. Karel. 1990. "Thermodynamics and kinetics of intracellular ice formation during freezing of biological cells." *J Appl Phys* 67 (3):1582–1593.

5

Storage and Shipping of Frozen Cells

Scientific Basis for Selection of a Storage Temperature

The selection of a temperature at which to store your cryopreserved cells is based upon scientific principles. Understanding these principles will allow for the proper selection of a storage temperature.

As described previously, cells are cryopreserved in order to prevent degradation of their critical biological properties. The use of low temperature to prevent degradation of biological systems reflects the desire to control water and its activity as well as the activity of degradative molecules present in the cell. The following is a brief overview of the physical behavior of these molecules at low temperature. More details can be found in Hubel, Spindler, and Skubitz (2014).

As described in Chapter 3, cryopreservation solutions contain an aqueous carrier solution, cryoprotective agents, and potentially proteins. In contrast to pure water, these solutions freeze over a range of temperatures. As the sample cools, ice forms in the extracellular space removing water in the form of ice (Figure 4.4). The remaining unfrozen solution is highly concentrated. As freezing progresses, water continues to be removed in the form of ice and the remaining unfrozen solution becomes more highly concentrated. The sample is not fully solidified until the sample forms a eutectic or a glass. The eutectic point is where the solidus and liquidus of a solution meet. A glass is a non-crystalline amorphous solid and is observed at temperatures below the glass transition temperature for a solution, T_g.

For a simple binary solution of NaCl and water, there is ice and unfrozen solution present in the sample until the sample reaches the eutectic temperature of $-21.2°C$. For cells preserved in solutions containing cryoprotective agents, empirical equations have been used to estimate T_g for multicomponent mixtures:

$$T_g\left(\text{mixtures}\right) = T_{g1} \cdot \left(1 - x\right) + T_{g2} \cdot x + k \cdot x \cdot \left(1 - x\right)$$

with the glass transition temperatures $T_{g,i}$ $(i = 1,2)$ of the components, the weight fraction, x, of the cryoprotective agent, and an interaction parameter,

Preservation of Cells: A Practical Manual, First Edition. Allison Hubel.
© 2018 John Wiley & Sons, Inc. Published 2018 by John Wiley & Sons, Inc.

k. The glass transition temperature of a 10% (w/w) dimethylsulfoxide solution is −132.58°C (Murthy 1998).

When the temperature of a solution goes below the glass transition temperature, the mobility of water molecules is reduced due to an increased viscosity of 10^{13} Pa·s (Angell 2002). Therefore, cells should be stored at temperatures below which the sample is fully solidified or vitrified, so that the water molecules in the sample are immobile and cannot participate in degradative processes.

All cells contain and secrete degradative molecules. These molecules participate in normal biological functions of the cells as well as their degradation. The activity of these molecules (most often proteins) is a function of temperature (McCammon and Harvey 1988). Reduced temperatures result in reduced protein dynamics/activity. Therefore, the reduction in protein activity with decreasing temperature is one mechanism by which cells are stabilized at low temperatures (specifically cryogenic temperatures).

The activity of proteins follows an Arrhenius relationship (Arrhenius 1889), whereby the rate constant, R, of a chemical reaction involving the protein is dependent on the absolute temperature, T:

$$R = A \cdot e^{-E_a/(R_u \cdot T)}$$

where A is a pre-exponential factor, E_a is the activation energy, and R_u is the universal gas constant. This relationship implies that the rate constant, R, decreases exponentially with decreasing temperature.

Therefore, to prevent further degradation, cells should be stored at a temperature where proteins present in the sample are no longer active. There are thousands of proteins present in a cell and clearly it would not be practical to measure the variation in activity with temperature of all of them. There have been limited studies to determine the behavior of proteins at low temperatures and these studies can help to rationally select a storage temperature below which protein activity is suppressed.

Decades ago, distinct changes in the dynamic properties of a variety of proteins were observed near −53°C (Bauminger et al. 1983, Doster, Cusack, and Petry 1989, Hartmann et al. 1982, Loncharich and Brooks 1990). Conventional wisdom at that time was that storage at temperatures below −53°C was sufficient to suppress protein activity. However, more recent studies demonstrated activity of proteins at temperatures far below −53°C.

Rasmussen and colleagues (1992) observed activity of ribonuclease A (RNA A) down to −58°C. Tilton Jr., Dewan, and Petsko (1992) observed changes in activity of RNA A at temperatures as low as −93°C. More and colleagues observed activity of β-glucosidase at temperatures as low as −70°C (More, Daniel, and Petach 1995). These studies demonstrate that for the limited number of proteins studied, protein activity may persist at very low temperatures (less than −80°C) and as a result, cells are now stored at lower temperatures (LN_2).

Additional Considerations for Vitrified Samples

The stability of vitrified samples is an important issue, but there is limited scientific literature on this topic. As described in Chapter 4, crack formation can be observed in vitrified samples cooled to temperatures near T_g and the risk of crack formation increases with decreasing temperature. As a result, it is common for vitrified samples to be stored in the vapor phase of liquid nitrogen (LN_2) at temperatures above that of LN_2 ($-196°C$) and only slightly below T_g. A limited number of studies suggest that storage of a sample at $13–14°C$ below T_g is recommended (see Fahy and Wowk 2015, for review). It is noteworthy that Mehl and colleagues observed that the critical warming rate to prevent extensive devitrification of a vitrified sample increased dramatically with time in storage (Mehl 1993). However, other studies have demonstrated the stability of red blood cells (RBCs) stored at $-80°C$ for up to 37 years in a solution containing 40% glycerol (Valeri et al. 2000). Storage temperature and changes in the sample with duration of storage are important issues for vitrified samples and additional studies may help guide protocol development and use.

This section was intended to provide an overview for the scientific selection of a storage temperature. A variety of studies have been performed measuring the stability of cryopreserved/vitrified cells and tissue with duration in storage and storage temperature (Fahy and Wowk 2015, Hubel, Spindler, and Skubitz 2014).

Standards, Guidelines, and Best Practices

Cryogenic storage of cells takes place in a repository. A repository may consist of anything from a single small storage Dewar in a laboratory to a free-standing repository containing large-storage Dewars with a bulk LN_2 storage and extensive monitoring and alarm systems. There are guidelines for the design and operation of repositories that store cryopreserved samples. The American Association of Tissue Banks, American Association of Blood Banks (AABB), Foundation for the Accreditation of Cellular Therapy (FACT), and International Society for Biological and Environmental Repositories (ISBER) all have guidelines and best practices for operating a repository. The best practices and guidelines from ISBER and FACT help to structure a repository and its operation (FACT 2015, ISBER 2012). *Conformance with the best practices alone is not sufficient.* It is possible to create a process that is compliant but a failure from a quality perspective. Constructing a repository that conforms to the scientific principles described above and best practices will result in the best outcome. Suggestions for integrating scientific principles with best practices are interspersed in this chapter.

Facilities

Repositories should have heating and air conditioning sufficient to maintain a reasonable temperature in the facility. If using mechanical refrigeration for storage of samples, sufficient air conditioning capacity will be required as these devices produce a large amount of excess heat. In general, temperatures above 22°C should not be permitted.

Air circulation should be sufficient to prevent excess humidity and condensation. Air filters should be cleaned or replaced regularly. Excess humidity can result in growth of fungus, which can represent a hazard for both staff and contamination of samples being stored. Mechanical freezers must be located at a proper distance from the wall to encourage air circulation and prevent excess heat build-up, which can reduce the lifespan of a compressor. Facilities using dry ice or LN_2 should have adequate ventilation and monitoring to ensure that adequate oxygen levels are maintained for personnel. Access to a repository should be controlled. It is common for storage Dewars to be locked or contained in a locked room, thereby limiting access to the facility to those authorized to do so. Reducing temperature excursions that the samples experience may also involve limiting the frequency of accessing the repository and requiring anyone with access to have training on sample deposition or retrieval. Storage of high-value samples (such as clinical samples) may require a dedicated security system for securing and monitoring the repository that may include key card access or video monitoring.

All repositories should have an emergency/disaster plan. Weather and man-made disasters can strike and impede proper function of the repository. Core elements of a disaster plan include a clear specification of personnel, their responsibilities, and the manner by which they can be contacted. These plans should also include specifications for alternatives should the person designated for a given responsibility be unavailable. Contact information for technical support (e.g., equipment suppliers or repair technicians) should be easily accessible.

For repositories storing critical samples, a disaster plan may also involve splitting of samples and data to several sites. For example, a second sample and backup data could be stored at another location. In the case where there is power loss, repositories should have plans for backup power or delivery of LN_2. It is common for repositories to have bulk LN_2 tanks filled before a storm or additional fuel delivered for backup electricity generation.

A complete description of disaster planning is beyond the scope of this chapter. There is growing pressure on repositories to expand the scope and robustness of their disaster plans. Additional resources include ISBER Best Practices (Campbell et al. 2012) as well as accreditation organizations such as FACT or the College of American Pathologists. These resources should be consulted for up-to-date recommendations for disaster planning.

Storage Equipment and Environment

Cryopreserved cells are commonly stored in specially designed LN_2 storage units called Dewars or freezers. These units typically contain internal racks designed to hold bags in cassettes (see Chapter 3), boxes containing vials, or canisters containing straws. These units come in a variety of capacities and may be supplied either by freestanding LN_2 supply tanks or by a central bulk tank.

Storage Dewars may store samples in the liquid or vapor phase of LN_2. Samples stored in the liquid phase should be placed in overwraps in order to prevent infiltration of LN_2 into the container. Currently, the preferred method of storage for samples is in the vapor phase of LN_2, which has several advantages. It prevents cross-contamination of samples in storage (Tedder et al. 1995). Vapor-phase storage also prevents entry of LN_2 into containers, which can result in failure of the container upon warming. When LN_2 vaporizes upon heating, it increases in volume 700–800 times; when trapped in a vial or bag the container can rupture.

Several different types of Dewars may be present in a repository. There may be a quarantine Dewar for samples that have been received but have not completed testing for infectious/adventitious agents or have tested positive for infection. Samples may be moved from the quarantine Dewar to a working or long-term storage Dewar upon completion of testing. Working Dewars can be used for storage of sample for relatively short periods of time (weeks to months) and are accessed on a regular basis. Samples that may be stored for an extended period of time may be placed in a long-term storage Dewar that is accessed very infrequently. The use of designations for Dewars (quarantine, working, and long-term storage) helps retain stability of the samples that are being stored. Factors that reduce the stability of samples will be discussed in the following.

The repository should also include an adequate supply of LN_2. Storage Dewars should contain at least 3 days' capacity at normal usage. Liquid nitrogen levels in the Dewar as well as in the supply tank should be checked at least once a week and more often for critical samples. It is common for storage Dewars to have electronic monitors that give an electronic readout of the LN_2 level.

In certain situations, storage of cells may be in a mechanical freezer. Single-stage mechanical freezers typically operate at −80°C. RBCs cryopreserved in glycerol (40%) are stored at −80°C in a mechanical freezer and have a shelf-life of about 10 years. Two-stage mechanical freezers typically operate at −150°C. In certain circumstances, nucleated cells may be stored at these temperatures in a mechanical freezer.

As with LN_2 storage, storage of samples in mechanical freezers should include backup systems. As described earlier, backup electricity should be available in case of power interruption. Many mechanical freezers may have emergency backup systems that use liquid carbon dioxide or LN_2 to cool the contents of the freezer in case of an extended power loss. Adequate supplies of the liquids used in the backup system should be available at all times.

All repositories should have a plan for preventative maintenance and repair of critical equipment. Maintenance should be performed typically as per manufacturers' recommendations. Calibration should also be performed routinely, in particular, for temperature monitoring devices present in both LN_2 storage units and mechanical freezers. Proper function of equipment should be verified before initial use or after a unit has been repaired.

Routine maintenance may also include cleaning/decontamination of units and defrosting of mechanical freezers. Cleaning solutions should be chosen with care to prevent damage to the surfaces being cleaned while still effective on fungus or other contaminants. Similarly, ice removal should be performed in a manner that does not damage the surface of the unit.

All equipment will eventually wear out. Low-temperature environments are particularly hard on electronic components (most often monitoring and alarm systems). A plan for replacement or repair of critical components should be put in place and built into preventative maintenance plans. When purchasing new equipment, the lifespan of the equipment should also be considered.

When equipment failure happens, it will be important to have adequate backup capacity for temporary storage of samples. It is recommended that backup capacity be to the capacity of the largest single storage unit and should be held in reserve. A conservative approach to sizing backup capacity is to reserve 10% of the existing repository capacity for backup. This approach requires a considerable investment that may be justified based on the value of the samples being stored and the potential risks for storage. Repositories should also have a written procedure for transferring samples from a failed or malfunctioning unit and for the return of the samples after repair or replacement. As described previously, critical samples may be split and stored in different locations in order to prevent loss due to equipment failure or disaster.

Mapping Storage Devices and Setting Alarm Limits

Although the temperature inside a storage unit can vary spatially, most storage units have only one temperature probe for monitoring temperature in the unit and that probe is in a fixed location. It is important to map the temperature in the storage unit. This information can be used to decide upon the region of the unit where samples should be stored. For example, samples that should maintain a temperature less than −150°C (cell therapy products) for maximum stability should be stored only in the region of the Dewar that maintains the desired temperature.

The mapping of the unit can also be used to set alarm limits. Dewars may have limits on low or high levels of LN_2 or on high- or low-temperature limits. Once again, mapping of the unit will enable selection of high and low limits based on maintaining the desired temperature in the space where samples are stored. It is common for the temperature at the top of a Dewar to be higher than that at the bottom. The high-temperature limit for the Dewar can be set such that the samples stored in the upper portion of the Dewar do not exceed −150°C.

Monitoring Systems

Low temperature is an adverse environment, especially for electronics. Circuit boards or electronic components may need to be replaced on a routine basis. *The sensitivity of monitoring systems and their potential for failure has resulted in the common practice for a second independent monitoring system to be put in place.* For example, it is common for LN_2 levels to be measured manually on a regular basis and compared to measurements on the control panel. As described above, most Dewars include a temperature measurement probe at a fixed location. This temperature probe will be connected to an alarm on the unit and if the high temperature threshold is achieved, the alarm will go off. For high-value samples, it is also common for Dewars to have a second independent temperature sensor present in the storage unit and connected to a central alarm system. Should failure of the primary temperature probe (or monitoring system) take place, the secondary probe is present and can provide monitoring as well.

As with Dewars, mechanical freezers used for storage will also have temperature probes present inside the storage area. Temperature limits can be specified and connected to alarm systems based on a temperature map of the freezer. Secondary independent monitoring can be performed using a second independent temperature probe. An alternative is to monitor voltage supply to the compressor.

Safety

Working in a repository implies contact with low-temperature environments. Contact between skin and cryogenic fluid or metal objects cooled to cryogenic temperatures can result in contact burns and if the duration of exposure is long enough, frost bite. Liquid nitrogen has a low viscosity and can therefore penetrate porous or woven clothing more rapidly than water.

Proper safety protocols when working with cryogens involves (i) developing a safety and training program that minimizes exposure; (ii) wearing proper clothing (nonabsorbent gloves, eye/face protection, proper footwear); (iii) working with team members instead of working alone; and (iv) informing other staff members when personnel are working with cryogenic fluids.

As described earlier, vaporization of cryogenic liquids can displace oxygen in the environment and result in asphyxiation. Oxygen concentrations as low as 13% can be tolerated under normal atmospheric conditions. Oxygen sensors (either portable or fixed) should be available and ventilation systems should permit 1–2 volume air changes per minute. If space is confined or the volume of cryogen being handled is large, a harness for retrieval of personnel may be recommended or portable breathing units may be available.

Hypothermia is the third basic safety consideration for operation of a repository. Operation in a low-temperature environment may result in exposure to cold to the extent that the body cannot maintain normal body temperature. Hypothermia can affect judgment and reaction times. Additional clothing may be needed for personnel working in large refrigerated repositories. In addition, it is recommended to limit exposure time and rotate personnel working in a cold environment.

The aforementioned description is just a brief overview of common safety issues associated with low-temperature working environments. A complex, large repository will face additional concerns regarding safety issues associated with oxygen enrichment (condensing oxygen out of the air) and over pressurization. Additional guidance on safety in cryogenic environments can be obtained from professional societies, such as the Cryogenic Society of America.

Inventory Management System

Repositories should also include an inventory management system. Knowledge of sample identity and location is critical to downstream use. Other information can be linked to the sample identifier, such as the sample processing record (cell type, cell count, cryopreservation method), patient medical records, consent forms (in the case of human biospecimens), and so on.

Inventory systems permit the sample to be located within the repository. Storage Dewars may contain thousands of samples and a precise location will enable rapid retrieval of the sample and thereby minimize damage to "innocent" samples (i.e., samples that are near the sample being retrieved).

Stability in Storage

As described earlier, storage temperatures are selected to suppress the mobility/activity of water and degradative molecules in the sample. If all degradative processes are suppressed, samples stored under those conditions should theoretically have an indefinite shelf-life. In reality, a variety of factors influence the stability of samples even when they are stored at the appropriate temperature. This chapter will describe two different factors: temperature fluctuations and background ionizing radiation.

Temperature Fluctuations

Samples are preserved so that they can be used for downstream applications. As a result, repositories will be accessed to add or remove samples. For repositories that do not have robotic sample retrieval, retrieval of a sample typically requires opening of the repository and removal of a rack that contains the sample of interest. Most racks contain more than one sample. If the sample is a vial, vials are typically contained in boxes and racks typically contain multiple boxes. Removal of a rack may result in the removal of hundreds of samples in both the box containing the vial of interest and also the other vials in the other boxes contained in the rack.

The vial that is removed from a storage Dewar and placed in a LN_2 cooled transit carrier will experience a rapid temperature rise followed by a slow decline in temperature (Figure 5.1). The vials in the same box surrounding the one that is removed will also experience a temperature rise and fall, albeit typically less than that experienced by the vial which was removed. The samples removed from the repository and then replaced after retrieval of the sample of interest are referred to as "the innocents" and they are subjected to temperature cycles each time that the rack is removed.

As repositories can be accessed on a regular (perhaps even daily) basis, temperature cycling of samples in the repository that are not being used can result in degradation of the sample over time. A limited number of studies have been performed quantifying the influence of this type of thermal cycling on post-thaw recovery. Studies of both peripheral blood mononuclear cells as well as umbilical cord blood (UCB) have observed that temperature excursions increase the fraction of cells that are apoptotic post-thaw (Cosentino et al. 2007, Hubel et al. 2015) and suggest that damage to cells resulting from temperature cycling is manifest through apoptotic pathways.

A variety of methods can be used to reduce degradation of samples resulting from temperature cycling. First of all, access to the repository should be limited. Requests for samples should be pooled and access to the repository should be limited to reduce the number of times in a day per week that the repository is accessed. Secondly, individuals who access the repository should be properly trained on sample retrieval and minimizing temperature excursions for the "innocents." Finally, designating Dewars to be as either working or long term will help reduce degradation samples to be used in the near future should be placed in a working Dewar and archival samples that may not be used or may be used very infrequently should be placed in a long-term

(a)

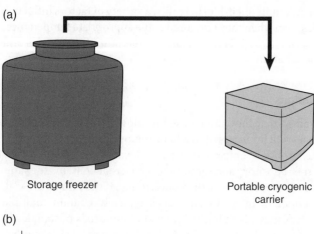

Storage freezer   Portable cryogenic
                                       carrier

(b)

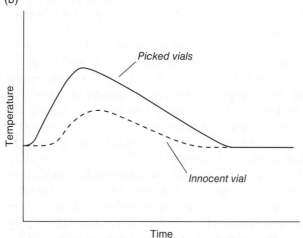

Figure 5.1 (a) Schematic representation of the transfer of a sample from a storage Dewar to a transport carrier; (b) temperature excursion of the sample (picked vial) being removed from the storage unit and placed in the transit carrier (solid line) as well as the "innocent vial" that was in the same rack and box as the picked vial (dashed line).

storage Dewar. Recently, automated storage retrieval systems have been developed and these systems are designed to reduce temperature excursions during sample retrieval.

Influence of Background Ionizing Radiation on Stability in Storage

Cryopreserved cells are sensitive to background ionizing radiation and there can be an accumulation of damage over time, in particular for long-term storage. The influence of ionizing radiation on the stability of frozen cells in long-term storage has been known since the mid-1950s (see Meryman 1966, for review). More recently, UCB units subjected to varying doses of radiation were studied and a distinct decrease in frequency of colony forming units (CFU) with increasing radiation dose was observed (Harris et al. 2010). As with temperature fluctuations, damage to cells resulting from ionizing radiation could be detected through an increase in apoptotic cells post-thaw (Cugia et al. 2011).

Shelf-Life of Samples in Storage

There is considerable interest in specifying the shelf-life of cells in long-term, low-temperature storage. For cells used therapeutically, regulatory agencies (e.g., the FDA) typically mandate an expiration date or specification of the shelf-life. The shelf-life of a limited number of cryopreserved cells has been determined. RBCs cryopreserved in 40% glycerol have a shelf-life of 10 years. Additional studies have demonstrated that RBCs that have been stored for up to 37 years can be thawed and recovered (Valeri et al. 2000). UCB has been collected and banked since the early 1990s. There is considerable interest in determining the shelf-life of cryopreserved UCB. *In vitro* studies have demonstrated that cord blood stored approximately 20 years can be thawed and cultured to form induced pluripotent stem cells (Broxmeyer et al. 2011). However, these studies do not demonstrate clinical activity of the cells, which is what would be needed to demonstrate that these cells stored for extended periods of time are still viable and functional. Even for cells that are not used therapeutically, there is a customer bias toward using cells that have not been in storage for extended periods of time.

Shelf-life studies are costly and time-consuming. A large number of samples need to be stored, thawed, and assayed over an extended period of time (a period of time beyond which the end user has hypothesized that the cells are stable). The difficulty in performing these studies means that they have not been performed on a wide range of products and public distribution of the studies that have been performed are very limited.

For pharmaceuticals, study methodologies have been developed to accelerate degradation of the sample and then use that data to determine a shelf-life of the product. There have been efforts to apply the paradigm developed for pharmaceuticals to cell preservation. At this point, there is no widely accepted method of performing accelerated degradation tests and the use of the test outcome to determine a shelf-life.

Fit-for-Purpose Storage Practices

As with other aspects of preservation, the storage practices used must reflect the downstream use of the sample. Critical parameters to consider include the duration of storage and the nature of the downstream use of the sample. If the storage of the sample is relatively short term (weeks to months), then the need to control the temperature of storage (and its variation with time or location) or the frequency of accessing the repository becomes less critical. If the sample is of high value, then monitoring the temperature of storage is critical as well as its variation with space and time. Limiting access to the repository containing the samples is also desirable which can be achieved by storage of samples in long-term storage Dewars that are not accessed routinely. Once again, these recommendations are based on the scientific understanding that the temperature of storage and its variation with time influences the stability of a sample.

Risk Mitigation in Long-Term Storage

The collection, processing, and preservation of cells can represent a significant investment. Certain cryopreserved cells (e.g., UCB, cell therapy products) cannot be replaced or can be replaced only at a significant cost. The banking of high-value cell products should also include a risk-management plan. A variety of threats may be present (e.g., severe weather, fire, terrorism, utility failure, etc.). There are core elements of a risk-mitigation plan. Many of these elements overlap with the conventional requirements of a repository and include the following:

- Two independent methods of identifying a sample
- Backup servers for information
- Splitting of critical samples
- Emergency plan and regular review
- Backup capacity
- Backup power/LN$_2$ supply
- Worker safety: oxygen measurement, safety plan
- Monitoring/alarm system for temperature monitoring of the storage units

Shipping or Transport of Cells

Cryopreserved cells are routinely shipped among sites of collection, processing, storage, and use. The transportation of cells may also cross national boundaries. The National Marrow Donor Program in the United States has international collection sites in Germany, Israel, Norway, Poland, Sweden, and the United Kingdom. As a result, hematopoietic stem cells donated in one country may be shipped to another country to a recipient that is the best match.

The critical biological properties of cells must be maintained during the transport or shipping processes. There are some subtle but critical difference between "transport" and "shipping." Transport is defined as the movement of a product within or between facilities. The product does not, however, leave the control of trained personnel at the originating or receiving facilities.

In contrast, shipping involves the transport of a product within or between facilities by a third party who is not a highly trained personnel for either the originating or the receiving facilities. For example, the transport of cryopreserved cells by a commercial package delivery service is considered shipping.

General Shipping Considerations

Classifying the sample being shipped is a critical first step in determining the regulatory requirements for the sample. Cryopreserved cells may be classified as dangerous goods and as a result, individuals who handle those goods must have special training.

A second step is to determine the specification of the shipping temperature. In general, cryopreserved cells are shipped at LN_2 temperatures in LN_2 dry shippers. Standards for shipping temperatures have been established by AABB, The United States Pharmacopeia (USP), and FACT. The temperature for storage or shipping of cell lines or cells being banked for nontherapeutic purposes should be monitored and should not exceed $-130°C$. Cell therapy units should not become warmer than $-150°C$ during shipping (USP 2014).

The quantity of material to be shipped should also be specified, as it will affect the type of packaging, the amount of refrigerant required, and the size of the shipping container. The size of shipping containers may be a significant constraint and may force division of the shipment into smaller volumes.

All shipping procedures and policies should be designed to protect both the quality/integrity of the sample and the health and safety of all personnel involved in the shipping process. General safety considerations have been described previously in this chapter.

Development of a shipping protocol typically involves verification, which should demonstrate that the shipping process results in acceptable post-thaw recovery. The process of verification can also provide helpful guidance when shipments deviate from the typical method or time frame. For example, weather can delay arrival of a sample and the verification process can be used to develop a window of acceptable shipping times before the sample could be considered compromised.

The verification process involves sending test shipments and measuring the ability to maintain the desired temperature range for times that extend beyond the normal range of time that would be expected and for a wide range of environmental conditions (high summer and low winter temperatures). Other factors that can be measured include (i) weight of the container before and after shipping as a measure of LN_2 consumption and (ii) resistance of the shipping unit to the physical stresses of shipment. Most verification studies do not involve shipment of the actual cell product. One option is to ship a cryopreserved cell line or cell type that is easily obtained or assayed to determine the influence of the test shipment on cell viability.

Shipping of a sample must confirm to all governing regulations. When traveling by air, shipments must conform to regulations of the Federal Aviation Administration, the International Civil Aviation Organization (ICAO), and the International Air Transport Association (IATA) standards. The ICAO is an agency in the United Nations that sets standards and regulations for international civil aviation. In contrast, the IATA is a trade association that develops regulations regarding the air transport of dangerous goods.

Ground shipments should conform to the applicable national standards. In the United States, those shipments must conform to the Department of Transportation (DOT) Standards. Packages also should meet other regulatory requirements for quarantine, biosafety, and biosecurity.

Liquid Nitrogen Dry Shippers

Cryopreserved cells typically are shipped in LN_2 vapor shippers (also known as dry vapor shippers) designed to keep the frozen sample cold. Dry shippers consist of a conventional cryogenic vessel that contains a porous material capable of absorbing LN_2. The product is contained in a cavity inside of the porous material. The use of a material capable of absorbing LN_2 eliminates the risks of contact with LN_2. There are several designs for vapor shippers and different designs will have different time periods over which the temperature will remain stable (typically between 5 and 20 days). The shipper should also remain cold for a period of time beyond the expected arrival time at the receiving facility (typically more than 48 h). The time buffer provides for the product to remain cold should there be delays in shipping or to store the product briefly in the shipper at the receiving site.

The stability of a dry shipper is strongly influenced by the orientation of the device. Dry shippers that are kept upright during the shipping process have much greater temperature stability than shippers that tipped on their side. Ratings from manufacturers for the temperature stability of shippers are based on the behavior of the unit in the upright position. A common method for insuring that the unit remains upright during shipment is to place the dry shipper in an outer container that is designed to keep the shipper in an upright position. The outer container may also contain padding that can reduce the vibration of the inner dry shipper and acts as a secondary containment vessel should there be any leakage from the inner vessel. Accredited facilities are required to ship products using both primary (dry shipper) and secondary containers (outer containers). Shipping units are subjected to significant vibrations and mechanical stresses during shipment and should be inspected on a regular basis for proper function and replaced regularly. The ability of the shipper to retain cryogen should be verified before each shipment; a weight loss over a 24-h period from a charged shipper is sufficient to calculate hold time and verify suitability for shipment.

Temperature Mapping of a Shipper

As with storage units, the temperature profile inside a shipper should be mapped and this mapping should be used to position the sample inside the unit and determine the duration of storage for the sample.

Packaging of Samples Being Shipped

The packing for the shipment should be able to withstand pressure changes, shocks, vibrations, temperature changes, punctures, and other conditions that would be considered routine during shipment. The dry shipper and secondary containment are considered part of the packaging system for the sample. Interior packaging of the sample will typically include a secondary container that is sealed to prevent leakage should the primary container be breached during shipment. In addition, interior packaging will be used to position the sample between refrigerant or cool surfaces (versus on top or on bottom of refrigerant), and pad the sample to prevent motion of the sample within the shipping container.

Monitoring of Shipments

As the temperature is a critical parameter, data loggers may be used to monitor temperature during shipping. Different models of loggers are available with certain units measuring the internal shipper temperature and other units capable of measuring both the ambient temperature and the internal shipper temperature. The temperature record can be combined with the shipping records to indicate whether or not the vessel was opened to retrieve the sample at the recipient facility or whether the vessel was tampered with during shipping. The record of temperature during transport can be downloaded and included as a part of the record for the sample.

Responsibilities

Three parties are typically involved when a sample is shipped: originator, courier, and recipient. Proper shipping procedures require involvement of all the three parties from the point where the sample leaves the originating site through when the sample is received by the recipient.

Originator is responsible for the following:

- Shipping requirements and the hazards associated with the sample and its shipment
- Classifying the sample as hazardous or nonhazardous
- Selecting a qualified courier
- Proper packaging and labeling of the sample
- Labeling the outer container to identify the sample and indicate specific shipping instructions

- Coordinating shipping with the courier and the receiving facility
- Preparation of the required documentation (including customs forms)
- Verifying of the sample identity prior to shipment
- Reviewing the final sample record to qualify the sample as suitable for shipment
- Qualifying the LN_2 dry shipper and its ability to maintain temperature
- Preparing the dry shipper as per manufacturer's instructions
- Developing the emergency procedures for the sample in case of accident

Courier is responsible for the following:

- Ensuring that shipping procedures conform to relevant regulations (i.e., airline regulations and security procedures, or customs or procedures)
- Inspecting the package and documentation upon receipt to determine whether or not the package is properly classified and prepared for shipment
- Proper handling of the package during shipment (i.e., orienting the container properly, reducing impacts, vibrations, etc.)
- Delivering the package and relevant documentation to the recipient
- Notifying the recipient of the scheduled date and time of arrival of the shipment
- Providing and requesting 24-h contact information for dangerous products

Receiving facility is responsible for the following:

- Documenting the condition of the shipment upon receipt
- Reporting condition issues (when present) to the courier
- Ensuring proper storage conditions for the shipment
- Returning the dry shipper to the originating site promptly
- Acknowledging receipt of the shipment to the originating site
- Documenting all the steps listed above

Sample Annotation

Samples that have been stored and shipped should have specific information added to the record of the samples' processing. The temperature and duration of storage should be specified. If the sample is shipped and the data logged, the temperature history of the sample and duration of shipping should also be noted in the sample record.

Scientific Principles

- All samples should be stored at a temperature at which all the water in the sample is immobilized (frozen or vitrified) and below which the degradative molecules present in the sample are inactive.
- Fluctuations in temperature reduce the shelf-life of a sample being stored.

Putting Principles into Practice

- The storage process should reflect the downstream use of the cells and the desired duration of storage.
- Best practices recommend that cells be stored at temperatures less than −130°C (less than −150°C for cell therapy products).
- Storage units should be temperature mapped and samples should be stored in the region where the unit maintains the desired temperature.
- Vitrified samples should be stored at temperatures near T_g for the solution. Storage stability remains an issue for vitrified systems.
- Monitoring systems for conditions inside the storage Dewar are important and should reflect the value of the samples being stored.
- All personnel should be trained on hazards. Proper procedures for a repository and proper safety equipment should be available.
- An inventory management system must be present to enable access and use of samples in storage.
- Shipping requires cooperation and communication amongst originating site, courier, and receiving facility.
- Shipping protocols should be verified.
- Shipping procedures should strive to maintain both the sample quality and also safety of all personnel involved in the shipping.
- During shipment use of temperature monitors is recommended.

References

Angell, C. A. 2002. "Liquid fragility and the glass transition in water and aqueous solutions." *Chem Rev* 102 (8):2627–2650.

Arrhenius, S. 1889. "On the reaction rate of the inversion of non-refined sugar upon souring." *Z Phys Chem* 4:226–248.

Bauminger, E. R., S. G. Cohen, I. Nowik, S. Ofer, and J. Yariv. 1983. "Dynamics of heme iron in crystals of metmyoglobin and deoxymyoglobin." *Proc Natl Acad Sci* 80 (3):736–740.

Broxmeyer, H. E., M. R. Lee, G. Hangoc, S. Cooper, N. Prasain, Y. J. Kim, C. Mallett, Z. Ye, S. Witting, K. Cornetta, L. Cheng, and M. C. Yoder. 2011. "Hematopoietic stem/progenitor cells, generation of induced pluripotent stem cells, and isolation of endothelial progenitors from 21- to 23.5-year cryopreserved cord blood." *Blood* 117 (18):4773–4777.

Campbell, L. D., F. Betsou, D. L. Garcia, J. G. Giri, K. E. Pitt, R. S. Pugh, K. C. Sexton, A. P. Skubitz, and S. B. Somiari. 2012. "Development of the ISBER best practices for repositories: collection, storage, retrieval and distribution of biological materials for research." *Biopreserv Biobank* 10 (2):232–233.

Cosentino, M., W. Corwin, J. G. Baust, N. Diaz-Mayoral, H. Cooley, W. Shao, R. Van Buskirk, and J. G. Baust. 2007. "Preliminary report: evaluation of storage conditions and cryococktails during peripheral blood mononuclear cell cryopreservation." *Cell Preserv Technol* 4:189–204.

Cugia, G., F. Centis, G. Del Zotto, A. Lucarini, E. Argazzi, G. Zini, M. Valentini, M. Bono, F. Picardi, S. Stramigioli, W. Cesarini, and L. Zamai. 2011. "High survival of frozen cells irradiated with gamma radiation." *Radiat Prot Dosimetry* 143 (2–4):237–240.

Doster, W., S. Cusack, and W. Petry. 1989. "Dynamical transition of myoglobin revealed by inelastic neutron scattering." *Nature* 337 (6209):754–756.

FACT. 2015. Common Standards for Cellular Therapies. Lincoln, NE: FACT.

Fahy, G. M., and B. Wowk. 2015. "Principles of cryopreservation by vitrification." *Methods Mol Biol* 1257:21–82.

Harris, D. T., J. Wang, X. He, S. C. Brett, M. E. Moore, and H. Brown. 2010. "Studies on practical issues for cord blood banking: effect of ionizing radiation and cryopreservation variables." *Open Stem Cell J* 2:37–44.

Hartmann, H., F. Parak, W. Steigemann, G. A. Petsko, D. Ringe Ponzi, and H. Frauenfelder. 1982. "Conformational substates in a protein: structure and dynamics of metmyoglobin at 80 K." *Proc Natl Acad Sci* 79 (16):4967–4971.

Hubel, A., R. Spindler, and A. P. Skubitz. 2014. "Storage of human biospecimens: selection of the optimal storage temperature." *Biopreserv Biobank* 12 (3):165–175.

Hubel, A., R. Spindler, J. M. Curtsinger, B. Lindgren, S. Wiederoder, and D. H. McKenna. 2015. "Postthaw characterization of umbilical cord blood: markers of storage lesion." *Transfusion* 55 (5):1033–1039.

ISBER. 2012. "2012 Best practices for repositories." *Biopreserv Biobank* 10 (2):81–161.

Loncharich, R. J., and B. R. Brooks. 1990. "Temperature dependence of dynamics of hydrated myoglobin: comparison of force field calculations with neutron scattering data." *J Mol Biol* 215 (3):439–455.

McCammon, J. A., and S. C. Harvey. 1988. Dynamics of Proteins and Nucleic Acids. New York: Cambridge University Press.

Mehl, P. M. 1993. "Nucleation and crystal growth in a vitrification solution tested for organ cryopreservation by vitrification." *Cryobiology* 30 (5):509–518.

Meryman, H. T. 1966. "Review of Biological Freezing." In Cryobiology, edited by H. T. Meryman, 1–114. New York: Academic Press.

More, N., R. M. Daniel, and H. H. Petach. 1995. "The effect of low temperatures on enzyme activity." *Biochem J* 305:17–20.

Murthy, S. S. N. 1998. "Some insight into the physical basis of the cryoprotective action of dimethyl sulfoxide and ethylene glycol." *Cryobiology* 36 (2):84–96.

Rasmussen, B. F., A. M. Stock, D. Ringe, and G. A. Petsko. 1992. "Crystalline ribonuclease A loses function below the dynamical transition at 220 K." *Nature* 357 (6377):423–424.

Tedder, R. S., M. A. Zuckerman, A. H. Goldstone, A. E. Hawkins, A. Fielding, E. M. Briggs, D. Irwin, S. Blair, A. M. Gorman, K. G. Patterson, and et al. 1995. "Hepatitis B transmission from contaminated cryopreservation tank." *Lancet* 346 (8968):137–140.

Tilton Jr., R. F., J. C. Dewan, and G. A. Petsko. 1992. "Effects of temperature on protein structure and dynamics: X-ray crystallographic studies of the protein ribonuclease-A at nine different temperatures from 98 to 320 K." *Biochemistry* 31 (9):2469–2481.

USP. 2014. "US Pharmacopeia-National Formulary Standards for Biologics." In Cryopreservation of Cells, vol. 1044, 1–31. Washington, DC: US Pharmacopeia.

Valeri, C. R., G. Ragno, L. E. Pivacek, G. P. Cassidy, R. Srey, M. Hansson-Wicher, and M. E. Leavy. 2000. "An experiment with glycerol-frozen red blood cells stored at −80 degrees C for up to 37 years." *Vox Sang* 79 (3):168–174.

6

Thawing and Post-Thaw Processing

As described in the previous chapters, the purpose of cryopreservation is to preserve the critical biological properties of a cell until the time and location of its downstream use have been determined. Taking a frozen sample out of a repository and using it for downstream applications require thawing the sample (also known as rewarming). The thawing process is therefore defined as taking the sample from a storage temperature and warming it to the temperature required for the downstream use.

In Chapter 4, the freezing process and its importance to the post-thaw viability of cells were described. A variety of chemical, mechanical, and thermal influences are present and will affect cell viability. Briefly, cells are sequestered into channels of unfrozen solution between adjacent ice crystals and therefore are subjected to high solute concentrations and mechanical forces, as water is removed in the form of ice during freezing. Reduced temperatures influence cell functions (e.g., metabolism, transport, etc.) and the cell responds to these changes by surviving or dying. The same chemical, thermal, and mechanical processes present during freezing are also present during thawing.

The thawing rate associated with the highest post-thaw survival is a function of both the solution composition and the cooling rate (Mazur 2004) of the cells. These two factors influence the amount of ice present and the size of the ice crystals (both inside and outside of the cell) and therefore will influence the nucleation and growth of ice crystals during warming (Karlsson 2001).

Slow frozen samples: Avoiding damage to the cells during thawing for most conventionally, slow-cooled cells requires rapid thawing rates (greater than 60°C/min).

Vitrified samples: During cooling of a vitrified sample, a small number of ice nuclei can form until the sample reaches the glass transition temperature, T_g. When warmed, those nuclei present can continue to grow and new nuclei can form and grow as well. Therefore, ice nucleation and crystal growth can be an important issue during warming of a vitrified sample. The nucleation and growth of ice crystals during warming is called devitrification and this process can produce opacification of the sample. Vitrified samples are typically clear

Preservation of Cells: A Practical Manual, First Edition. Allison Hubel.
© 2018 John Wiley & Sons, Inc. Published 2018 by John Wiley & Sons, Inc.

and devitrification can result in clouding of the sample. Devitrification can produce the same changes in extracellular solute concentration and sequestration of cells into gaps between adjacent ice crystals that result during conventional freezing, which is undesirable. In order to avoid devitrification, the warming rate necessary to avoid nucleation and growth of ice during thawing is typically *two orders of magnitude greater than the cooling rate used for the protocol* (Hopkins et al. 2012). The combination of vitrification solution composition and cooling rate can result in the formation of stable, metastable, or unstable glasses.

Unstable glasses form at relatively low concentrations of cryoprotective agents. On a bulk level, unstable glasses do not exhibit crystallization but have been shown to contain very small ice nuclei that will grow rapidly during warming. Higher concentration regions are associated with the formation of metastable glasses. These compositions can form bulk glasses without a detectable freezing event, and devitrification may be suppressed with a sufficiently high warming rate. For even higher cryoprotectant concentrations, glasses are stable and devitrification will not be observed even for slow-cooling rates. Samples that have been vitrified to form metastable or unstable glasses require higher warming rates than those required for stable glasses (Fahy et al. 1984).

The warming rate has to be uniform across the specimen in order to reduce the thermo mechanical stresses, which can result in cracks forming in the sample. Crack formation is especially damaging when multicellular systems, such as tissues or organs, are vitrified. Vitrification of organs and tissues are active areas of research. Currently, vitrification of multicellular systems is not routinely performed clinically or commercially.

Thawing Equipment

Warm water bath: The most common equipment used to thaw vials or bags of cells are warm water baths. Samples can be removed from the repository/ freezer, immersed in the water bath, and swirled until the sample is thawed. Vials that are frozen are typically immersed up to the cap in order to prevent bath water from entering into the sample. Commercially available vial holders enable thawing of multiple vials at the same time. Warm water baths used for sample thawing are typically set at 37°C.

After thawing the cells, the outer surface of the vial or bag should be cleaned. It is common to use an alcohol wipe to clean the outer surface of the container that has been thawed in a water bath to reduce the potential for contamination when it is used for downstream applications.

The use of a water bath for thawing has several limitations. Operator-to-operator variability in the thawing process can be significant. It is not uncommon for a rack to be placed in a water bath and a vial simply placed in the bath to thaw. If the sample is not swirled, it reduces the warming rate. The operator may even continue to pick other samples from the repository and thaw them, resulting in variable thawing rates and variable processing times for the samples.

The water bath is open to the environment, which means that the process can result in contamination of the outer surface of the container. The potential for contamination is particularly concerning for cell therapies thawed in a water bath.

Controlled thawing devices: The limitations listed earlier associated with thawing of samples using a water bath have motivated development of alternative methods of thawing samples. Thawing devices have recently been developed that can be used for a variety of formats (vial or bag). These devices use electric heaters to thaw the sample, and the thawing process is meant to mimic that of a water bath. Current thawing devices that are only now becoming commercially available represent an improvement over the warm water bath. It is likely that with time these devices will become the predominant method of thawing, in particular for high-value samples.

Transporting Samples Prior to Thawing

Samples are stored in a repository but may be thawed at the site of use. For example, it is common for hematopoietic stem cells to be thawed at the patient's bedside and infused directly. Vials of banked cells are commonly thawed near a biological safety cabinet, so that they can be seeded into culture vessels. As a result, it is common for samples to be removed from the repository and transported to this site of post-thaw use where they are thawed.

As described in Chapter 4, transport of the sample from one site to the next during process (in contrast with shipping) requires use of the proper sample carrier. If the sample is stored in liquid nitrogen (LN_2), the sample carrier should maintain that temperature during transport from the repository to the site of thawing. There are commercially available transport devices capable of maintaining the proper temperature for transport of the sample. Transporting the sample on dry ice implies that the sample has been rapidly warmed from −196 to −80°C and then held there until thawing commences. This type of temperature excursion can influence post-thaw recovery.

Estimating Your Thawing Rate

Regardless of the method used to thaw the cells, the thawing process should be characterized. One method of estimating the thawing rate involves placing a thermocouple in the container (e.g., straw, vial, or bag) containing the cryopreservation solution to be used and then cooling the container using the desired method to the desired storage temperature. The sample is then removed and placed in the warming bath and thawed with the temperature as a function of time recorded. The average thawing rate, B_w, can then be approximated:

$$B_w = \frac{\left(T_{final} - T_{initial}\right)}{\left(t_{final} - t_{initial}\right)}$$

where T_{final} is the temperature of the sample at the end of thawing. $T_{initial}$ is the initial temperature (ideally, the temperature of the repository or transport vessel). The initial time, $t_{initial}$ is set at zero and the final time, t_{final}, is the time associated with the end of thawing. There is no standard definition for the end of thawing. However, it is common for samples to be thawed until there is just a small ice crystal remaining in the sample and the remaining sample is liquid. This approach prevents over heating of the sample. If the end of thawing is associated with the melting of the last bits of ice, t_{final} will be at the end of the temperature plateau associated with the latent heat of melting (see Figure 6.1) and that temperature should be the melting temperature of the solution. For an isotonic saline solution, the melting temperature is roughly –0.5°C and a 10% dimethylsulfoxide (DMSO) solution has a melting temperature of roughly –1.5°C.

If the thawing rate is below the desired threshold level then there are two basic strategies for increasing the thawing rate (i.e., increasing the heat transfer rate): (i) increasing the temperature difference (e.g., increasing the bath temperature) and (ii) increasing the heat transfer coefficient by swirling the sample in the bath. The water temperature of the bath could be increased to 42–45°C.

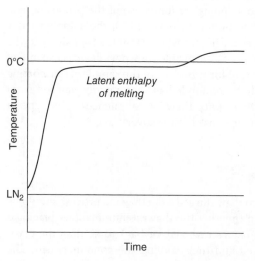

Figure 6.1 Typical thawing curve for a sample. Initial warming rates will be high and the warming rate will diminish as the sample approaches the melting temperature of the solution. The duration of the plateau associated with the latent enthalpy of melting will vary with the volume of the solution and the heat transfer rate of the sample.

Using a higher bath temperature brings with it the risk of *hyperthermic* damage to the cells; in particular, if a bath temperature of 45°C is used, which is the lower threshold temperature for hyperthermic damage. Swirling or mixing can be used to increase the heat transfer from the surface of the container.

Samples that contain a large volume or are thick will be difficult to warm rapidly. Cells frozen in bags are most commonly frozen in metal cassettes. The purpose of the cassette is to create a uniform thickness of the sample during freezing, resulting in more uniform heat transfer (and hence freezing). Bags used for cryopreservation typically specify a maximum volume of liquid that can be frozen in order to produce a thin but uniform layer after the bag is placed in a press. The uniform thickness of the frozen bag helps make the cooling more uniform during freezing and the warming more uniform during thawing.

It is common for multiple samples to be thawed at once. As discussed previously, thawing of multiple samples can lead to variations in the warming rate during thawing or also variations in the time between thawing and downstream processing of the sample. Cells are sensitive to DMSO post–thaw, and delays in washing of the cells can result in losses post-thaw from biochemical toxicity.

Overwraps may be used for either vials or bags to prevent loss of product should the bag break or to prevent influx of LN_2 into the sample. The presence of an overwrap may slow warming of a sample. If the sample is contained in an overwrap, additional characterization of warming should be performed in order to ensure that the warming rate is sufficient.

Thawing and Infusion of Cell Therapy Products

There are special considerations for thawing of cell therapy products. The following is a brief overview of these considerations. Details addressing infusion-related adverse reactions are reported in Haspel (2016).

If multiple bags are to be infused, each bag should be thawed and infused separately in order to monitor for infusion-related adverse reactions. Bags are typically frozen in cassettes, which are to be removed before the bag is thawed. Care must be taken when removing the bag from the cassette since the bag and the accompanying tubing will be cold and may be brittle. Rough handling can result in cracking of the bags or tubing and therefore in potential loss of the sample.

Concerns over bag failure have led to the practice of placing bags in an overwrap prior to thawing. Bag failure is relatively rare (1–4% depending upon the study) (Khuu et al. 2002), but cell therapy products are typically expensive or irreplaceable. Cracks or tears in the bags or tubing should be reported, and follow-up sterility testing or prophylactic treatment of the patient may be warranted.

Products that have been thawed should be inspected for clumping or aggregation. Cell losses from freezing can result in clumping or aggregation. Standard blood filters have been used to prevent infusion of clumps. Supplementation of the product with DNAse has also been used to minimize clumping or aggregation.

Safety Considerations for Thawing

Thawing can be hazardous for samples that have been stored improperly. If a sample has been stored in the liquid phase of LN_2 without an overwrap, there may be LN_2 trapped inside the vial. As the vial is warmed during thawing, the LN_2 in the vial will evaporate, expand, and ultimately rupture the vial. The rupturing of the vial can be a safety hazard. Vials that may contain LN_2 cannot be used, so they could be thawed in a container capable of withstanding the vial rupture and then disposed of properly.

Post-Thaw Processing

As described in Chapter 3, conventional cryopreservation solutions are not physiological solutions. Typically, these solutions are removed or diluted prior to downstream use of the cells. Post-thaw processing may also involve post-thaw assessment or post-thaw characterization, which will be covered in Chapter 7.

Post-Thaw Washing

The most common method of post-thaw processing involves washing the cells after thawing and before downstream use. Centrifugation of the cells followed by removal of the supernatant and replacing the solution with either a wash solution or culture medium is the most common method of washing cells post-thaw. This process may be repeated in order to reduce residual levels of cryoprotectants below the desired level. Manual washing of the cells (removal and addition of solution are performed manually) is done most commonly. Automatic centrifuges are also commercially available.

Manual methods of post-thaw washing are labor intensive, require expensive capital equipment, and can result in high cell losses, in particular for cells processed in bags (Antonenas, Bradstock, and Shaw 2002). New approaches are being developed to enable rapid processing of cells post-thaw. A device for post-thaw washing of cells has been commercialized consisting of a spinning membrane to concentrate cells after being diluted. Other devices for post-thaw washing are in development including a hollow fiber bioreactor (Zhou et al. 2011), a dead-end filtration device (Tostoes et al. 2016), and a microfluidic device (Hanna, Hubel, and Lemke 2012). These devices are intended to automate the washing process and thereby improve consistency and reduce cell losses.

Dilution

Dilution of the cells post-thaw is commonly performed for cells that are cultured downstream. The most common method involves taking thawed cells and adding them to culture media. The cells are cultured for about 24 h, at which time the media is removed and replaced, further diluting the cryopreservation solution present in the sample. The initial dilution of the sample immediately post-thaw should be sufficient to prevent toxicity from DMSO or transformation of the cells resulting from exposure to DMSO (typically 0.5% final concentration).

Infusion of Cells Immediately Post-Thaw

In certain contexts (e.g., hematopoietic progenitor cell transplants), cryopreserved cells are not washed post-thaw, but they are infused directly into a recipient. This approach can be considered a variation on post-thaw dilution, as the recipients blood is being used to dilute the cryoprotective agents and the patient will metabolize it (typically DMSO) to completely remove it. Minimally manipulated hematopoietic stem cells used for transplantation are the most common product handled in this manner. Conventional wisdom was that hematopoietic progenitor cells are sensitive to DMSO, in particular post-thaw, so the cells are thawed at the patient's bedside and infused immediately post-thaw in order to reduce cell losses.

If the cryopreservation solution is injected into the patient with the cells, the cryopreservation solution is considered an excipient. Cryoprotective agents used in cell therapy applications have included DMSO, glycerol, and propylene glycol. DMSO is not approved for infusion, and the administration of DMSO-containing cell suspensions are commonly associated with adverse events such as nausea, chills, hypotension, dyspnea, and cardiac arrhythmia. More serious reactions such as cardiac arrest, transient heart blockage, neurological toxicity, renal failure, and respiratory arrest have also been observed but far less often (Alessandrino et al. 1999, Benekli et al. 2000, Davis, Rowley, and Santos 1990, Galmes et al. 1996, Hoyt, Szer, and Grigg 2000, Martino et al. 1996, Stroncek et al. 1991, Syme et al. 2004, Zambelli et al. 1998, Zenhausern et al. 2000).

Historically, frozen and thawed cells containing DMSO, which is used therapeutically, have been infused into patients for one-time administration of a cell therapy to treat life-threatening diseases (e.g., bone marrow transplant to treat leukemia). New and emerging applications of cellular therapies may involve multiple treatments (e.g., immunotherapy) or the treatment of a non–life-threatening disease (e.g., autism, urinary incontinence, diabetes, etc.). In these contexts, it is unlikely that infusion of DMSO and the associated side effects would be considered acceptable. There is a movement toward washing of cells containing DMSO prior to infusion or the elimination of DMSO as a cryoprotectant for cell therapy products.

Glycerol has been used in red blood cell preservation since the 1970s. Infusion of glycerol-containing cells has not been associated with adverse events, and glycerol is considered safe for infusion up to certain levels. In contrast to DMSO, post-thaw processing of cells cryopreserved in glycerol to be given therapeutically must be washed, not because of toxicity issues but because of osmotic stress. For example, red blood cells cryopreserved in glycerol must be washed and the glycerol reduced to levels less than 3% vol/vol in order to prevent intravascular lysis of the cells.

Removal of Vitrification Solutions

As described in Chapter 3, vitrification solutions are high-concentration (4–6 M) solutions, and methods of removal/dilution can be as sophisticated as when they are introduced to the cells or tissues. It is not uncommon for removal of a vitrification solution to require two different wash solutions/steps. In general, cells will only tolerate a twofold change in osmolarity upon washing (versus fourfold for introduction). The density of a vitrification solution or even a wash solution can be comparable to or even greater than that of the biological system being vitrified. As a result, it is not possible to centrifuge the sample and separate the cells from the solution. Typically, tissues or gametes (oocytes or embryos) are vitrified, and techniques for transferring the sample between solutions have been developed (Fahy and Wowk 2015).

Wash Solutions

For low-concentration cryopreservation solutions or cells that are more tolerant of osmotic stress, wash solutions are isotonic (phosphate-buffered saline, tissue culture medium, etc.). In contrast to the introduction of a cryopreservation solution, transfer of a cell from a cryopreservation solution into an isotonic environment results in a rapid influx of water into the cell followed by a slow efflux of the penetrating cryoprotective agent (Figure 6.2a). Hematopoietic progenitor cells in umbilical cord blood are sensitive to osmotic stresses resulting from removal of DMSO, so a specialized wash solution has been developed (Rubinstein et al. 1995). It consists of 10% Dextran 40 in saline with 5% human serum albumin. This slightly hypertonic solution draws out DMSO from inside the cell without subjecting the cells to large volumetric excursions. Specifically, the Dextran 40

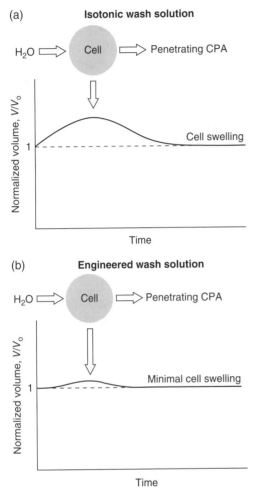

Figure 6.2 (a) Normalized cell volume as a function of time for cells washed in an isotonic saline solution. Water enters the cell to balance the high chemical potential followed by slower efflux of penetrating cryoprotective agent; (b) normalized cell volume as a function of time for cells washed in an engineered wash solution. The solution is slightly hypertonic and contains molecules that exert an osmotic force but are too large to penetrate the cell. Water influx is diminished when compared to part (a). Changes in cell volume are reduced.

exerts an osmotic force but does not penetrate the cell, so there is a driving force for DMSO to exit, but less of a driving force for water to leave the extracellular solution and enter the cell (Figure 6.2b). An engineered wash solution should reduce the osmotic stresses for cells being washed post-thaw and therefore minimize cell losses during this step. Ideally, the composition of the wash solution is such that an additional wash step is not required for downstream use of the cells.

For vitrification solutions, wash solutions will depend upon the composition and final concentration of the vitrification solution. Multistep protocols for removal are common post-thaw when samples are vitrified. For example, oocytes that are vitrified in 15% DMSO + 15% ethylene glycol are washed first in a 1 M sucrose solution, followed by a 0.5 M sucrose solution, before finally being resuspended in an isotonic media. As with conventionally cryopreserved cells, the wash solution is hypertonic and contains solutes that do not penetrate the cell membrane (e.g., sucrose). These solutions then draw out the penetrating cryoprotective agents (in this case, DMSO and ethylene glycol) while minimizing water efflux into the cell and large volumetric excursions. Wash solutions with three different compositions are required to bring the oocytes from the vitrification solution to isotonic conditions without significant losses.

Scientific Principles

- Rapid thawing is critical for both conventional slow-freezing protocols and vitrification. The thawing rate for vitrification would be at least one order of magnitude greater than the cooling rate in order to prevent devitrification.
- Removal/dilution of the cryopreservation solution is typically required before the cells can be used for downstream applications.

Putting Principles into Practice

- Estimating the average warming rate requires only a stop watch and the ability to observe the sample.
- New warming devices will improve the consistency and provide documentation as to the thawing process.
- Wash solutions for removal of cryopreservation solutions can be engineered to reduce osmotic stress and cell losses during washing.

References

Alessandrino, P., P. Bernasconi, D. Caldera, A. Colombo, M. Bonfichi, L. Malcovati, C. Klersy, G. Martinelli, M. Maiocchi, G. Pagnucco, M. Varettoni, C. Perotti, and C. Bernasconi. 1999. "Adverse events occurring during bone marrow or peripheral blood progenitor cell infusion: analysis of 126 cases." *Bone Marrow Transplant* 23 (6):533–537.

Antonenas, V., K. F. Bradstock, and P. J. Shaw. 2002. "Effect of washing procedures on unrelated cord blood units for transplantation in children and adults." *Cytotherapy* 4 (4):16.

Benekli, M., B. Anderson, D. Wentling, S. Bernstein, M. Czuczman, and P. McCarthy. 2000. "Severe respiratory depression after dimethylsulphoxide-containing autologous stem cell infusion in a patient with AL amyloidosis." *Bone Marrow Transplant* 25 (12):1299–1301.

Davis, J., S. D. Rowley, and G. W. Santos. 1990. "Toxicity of autologous bone marrow graft infusion." *Prog Clin Biol Res* 333:531–540.

Fahy, G. M., and B. Wowk. 2015. "Principles of cryopreservation by vitrification." *Methods Mol Biol* 1257:21–82.

Fahy, G. M., D. R. MacFarlane, C. A. Angell, and H. T. Meryman. 1984. "Vitrification as an approach to cryopreservation." *Cryobiology* 21 (4):407–426.

Galmes, A., J. Besalduch, J. Bargay, N. Matamoros, M. A. Duran, M. Morey, F. Alvarez, and M. Mascaro. 1996. "Cryopreservation of hematopoietic progenitor cells with 5-percent dimethyl sulfoxide at -80 degrees C without rate-controlled freezing." *Transfusion* 36 (9):794–797.

Hanna, J., A. Hubel, and E. Lemke. 2012. "Diffusion-based extraction of DMSO from a cell suspension in a three stream, vertical microchannel." *Biotechnol Bioeng* 109 (9):2316–2324.

Haspel, R. L. 2016. "Thawing and Infusing Cellular Therapy Products." In Cellular Therapy: Principles, Methods and Regulations, edited by E. M. Areman and K. Loper, 457–467. Bethesda, MD: AABB.

Hopkins, J. B., R. Badeau, M. Warkentin, and R. E. Thorne. 2012. "Effect of common cryoprotectants on critical warming rates and ice formation in aqueous solutions." *Cryobiology* 65 (3):169–178.

Hoyt, R., J. Szer, and A. Grigg. 2000. "Neurological events associated with the infusion of cryopreserved bone marrow and/or peripheral blood progenitor cells." *Bone Marrow Transplant* 25 (12):1285–1287.

Karlsson, J. O. 2001. "A theoretical model of intracellular devitrification." *Cryobiology* 42 (3):154–169.

Khuu, H. M., H. Cowley, V. David-Ocampo, C. S. Carter, C. Kasten-Sportes, A. S. Wayne, S. R. Solomon, M. R. Bishop, R. M. Childs, and E. J. Read. 2002. "Catastrophic failures of freezing bags for cellular therapy products: description, cause, and consequences." *Cytotherapy* 4 (6):539–549.

Martino, M., F. Morabito, G. Messina, G. Irrera, G. Pucci, and P. Iacopino. 1996. "Fractionated infusions of cryopreserved stem cells may prevent DMSO-induced major cardiac complications in graft recipients." *Haematologica* 81 (1):59–61.

Mazur, P. 2004. "Principles of Cryobiology." In Life in the Frozen State, edited by B. J. Fuller, N. Lane, and E. Benson, 3–66. Boca Raton, FL: CRC Press.

Rubinstein, P., L. Dobrila, R. E. Rosenfield, J. W. Adamson, G. Migliaccio, A. R. Migliaccio, P. E. Taylor, and C. E. Stevens. 1995. "Processing and cryopreservation of placental/umbilical cord blood for unrelated bone marrow reconstitution." *Proc Natl Acad Sci U S A* 92 (22):10119–10122.

Stroncek, D. F., S. K. Fautsch, L. C. Lasky, D. D. Hurd, N. K. Ramsay, and J. McCullough. 1991. "Adverse reactions in patients transfused with cryopreserved marrow." *Transfusion* 31 (6):521–526.

Syme, R., M. Bewick, D. Stewart, K. Porter, T. Chadderton, and S. Gluck. 2004. "The role of depletion of dimethyl sulfoxide before autografting: on hematologic recovery, side effects, and toxicity." *Biol Blood Marrow Transplant* 10 (2):135–141.

Tostoes, R., J. R. Dodgson, B. Weil, S. Gerontas, C. Mason, and F. Veraitch. 2016. "A novel filtration system for point of care washing of cellular therapy products." *J Tissue Eng Regen Med.* doi: 10.1002/term.2225.

Zambelli, A., G. Poggi, G. Da Prada, P. Pedrazzoli, A. Cuomo, D. Miotti, C. Perotti, P. Preti, and G. Robustelli della Cuna. 1998. "Clinical toxicity of cryopreserved circulating progenitor cells infusion." *Anticancer Res* 18 (6B):4705–4708.

Zenhausern, R., A. Tobler, L. Leoncini, O. M. Hess, and P. Ferrari. 2000. "Fatal cardiac arrhythmia after infusion of dimethyl sulfoxide-cryopreserved hematopoietic stem cells in a patient with severe primary cardiac amyloidosis and end-stage renal failure." *Ann Hematol* 79 (9):523–526.

Zhou, X., Z. Liu, Z. Shu, W. Ding, P. Du, J. Chung, C. Liu, S. Heimfeld, and D. Gao. 2011. "A dilution-filtration system for removing cryoprotective agents." *J Biomech Eng* 133 (2):021007.

7

Post-Thaw Assessment

The overall objective of a preservation protocol is to maintain the critical biological properties of a cell so that it can be used at a later time. What is meant by "critical biological properties" depends upon the cell type and its intended purpose. Therefore, post-thaw assessment must be fit-for-purpose. Different categories of post-thaw biological properties will be described below and the manner by which they are used to characterize a cell post-thaw will also be described.

Quantifying and characterizing the viability of a cell post-thaw is a critical step in preservation protocol. *Of all the steps of a preservation protocol, post-thaw assessment is the step that is done incorrectly most often.* Lack of an accurate and meaningful measure of post-thaw viability makes it impossible to validate a given preservation protocol.

Why is post-thaw assessment so difficult to perform well? Unlike a cell in culture or a cell that has been recently harvested from an *in vivo* environment, cells that have been frozen and thawed have been subjected to distinct, significant stresses. These stresses have altered the cell, such that methods used to quantify viability may not reflect the actual viability of the cell (Pegg 1989). Specifically, a cell may have dehydrated to a fraction of its initial volume during cryopreservation. After warming, the cell volume may have recovered, but changes in the cell membrane persist, and it takes time post-thaw for the membrane to recover.

Changes in the cytoskeleton have also been observed post-thaw (Chinnadurai et al. 2014), which in turn influence post-thaw function of the cells. Some of these alterations resolve with time post-thaw, implying that some cells may need to be cultured post-thaw before they can be used for downstream applications. As cells can experience post-thaw apoptosis, the viability of a cell population can vary with time post-thaw, thereby influencing the uses of the cells immediately post-thaw as well as the accuracy of post-thaw assessment.

Preservation of Cells: A Practical Manual, First Edition. Allison Hubel.
© 2018 John Wiley & Sons, Inc. Published 2018 by John Wiley & Sons, Inc.

Common Measures Used in Post-Thaw Assessment

A variety of methods can be used to characterize the post-thaw function of a cell. Mechanical integrity assays are commonly performed post-thaw. Additional assays include metabolic activity, mechanical activity, mitotic activities, differentiation potential, and transplantation potential. (See Table 7.1 for summary.) Finally, post-thaw assessment should reflect post-thaw uses. For cells to be used therapeutically, post-thaw function is required. Post-thaw measures should therefore reflect the intended downstream application and function.

Physical Integrity

It is common for the term "viability" to be used interchangeably with the physical integrity of a cell. The most common method of determining physical integrity is a membrane integrity dye. This measure is commonly used as it can be performed rapidly and with minimal equipment and training. Cells can be placed under a microscope and the physical integrity of the cells can be quantified. Bleb formation or membrane irregularities may be noted as well.

Commonly used membrane integrity dyes include trypan blue, which requires only conventional light microscopy, whereas a fluorescent microscope is needed to be used with the use of 7-aminoactinomycin, acridine orange, and

Table 7.1 Summary of different categories of post-thaw assessments for cells: Multiple measures of post-thaw assessment should be used to characterize the cells post-thaw.

Mechanical integrity	*Mechanical activity*
• Membrane integrity using dyes	• Attachment
• Membrane disruption	• Migration
Metabolic activity	• Phagocytosis
• Consumption of nutrient or metabolite	• Motility
• Detection of products of metabolism	• Contractility
Differentiation potential (Stem cells only)	• Aggregation/self-assembly
• Multilineage differentiation (i.e., CFU and trilineage assays)	*Mitotic activity*
• Teratoma formation	• Proliferation
	• Cell cycle analysis
	Transplantation
	• Syngenic
	• Xenogenic

propidium iodine. There are other types of membrane integrity dyes that can detect metabolic activity of the cell as they only fluoresce after they have been internalized and interact with specific substrates inside the cell that reflect metabolic activity. For example, Calcein AM is a cell-permeant dye that can be used to determine cell viability. The nonfluorescent Calcein AM is converted to a green-fluorescent calcein after acetoxymethyl ester hydrolysis by intracellular esterases.

Another approach to measure membrane integrity is to detect the release of intracellular enzymes into the extracellular media. Lactate dehydrogenase (LDH) is an intracellular enzyme that can be detected in the extracellular solution only upon lysing of the cell. A variety of commercial methods for measuring LDH are currently available. The quantity of LDH/cell can vary from cell to cell (or amongst cell types) and as a result, using LDH release quantitatively brings with it a large error.

Metabolic Activity

All cells must exhibit metabolic activity post-thaw to be considered viable. There are two basic methods of determining the metabolic activity of a cell: measuring consumption of a nutrient/metabolite or detection of a product of metabolism.

A common method of determining metabolic activity is by measuring oxygen consumption. Measuring oxygen consumption can be performed using a Clark electrode, electronic paramagnetic resonance oximetry, and fluorescent probes (Diepart et al. 2010). Each method has advantages and disadvantages in terms of cost, sample size required, and complexity. Oxygen consumption can vary significantly with time post-thaw, so measurements should be performed at a consistent time point.

Other methods of measuring metabolic activity post-thaw involve detecting activity of common metabolic pathways. Nicotinamide adenine dinucleotide phosphate (NADP) and its reduced form (NADPH) are involved in a variety of biochemical reactions, including oxidation-reduction for the protection of cells against reactive oxidative species, allowing the regeneration of reduced glutathione, and anabolic pathways. A common method of detecting NADPH involves the use of a tetrazolium dye, 3-(4,5-dimethylthiazol-2-yl)-2,5-diphenyltetrazolium bromide (MTT). The MTT dye is yellow and undergoes a color change to purple after reacting with a mitochondrial reductase. The absorbance of the solution can be quantified using a spectrophotometer and metabolic activity can be quantified based on the intensity of the absorbance. Several other tetrazolium salts (INT (2-(4-Iodophenyl)-3(4-nitro-phenyl-5-phenyl)-5-phenyl-2*H*-tetrazolum chloride), TTC (3-(4,5-dimethyl-thiazol-2-yl)-5-(3-carboxymethoxyphenyl)-2-(4-sulphenyl)-2*H*-tetrazolium), XTT

(2,3-bis-(2-methoxy-4-nitro-5-sulfophenyl)-2*H*-tetrazolium-5-carboxanilide), and MTS (3-(4,5-dimethylthiazol-2-yl)-5-(3-carboxymethoxyphenyl)-2-(4-sulfophenyl)-2*H*-tetrazolium)) are commercially available and can be used in a similar fashion to MTT.

For certain cell types, specific synthetic or metabolic activities are critical for downstream use and therefore post-thaw assessment may include assays of this activity. For mesenchymal lineage cells, production of the proper extracellular matrix is critical for post-thaw function. The uptake of radioactive precursors for collagen can be used as a measure of the production of an extracellular matrix. Cultured cells may also be fixed and then stained for the presence of collagen using conventional histological dyes, such as Sirius red or Masson's trichome staining (Segnani et al. 2015).

Proper function of other cell types may require synthesis and secretion of soluble factors. Production of albumin has been used to characterize the metabolic activity of hepatocytes (Dunn et al. 1989). Alkaline phosphatase (ALP) is secreted by all cells, but it is found in greater abundance on the cell membrane in undifferentiated cells. It is common to stain embryonic stem (ES) or induced pluripotent stem (iPS) cells for the presence of ALP as a marker of cell primitivity (Thomson et al. 1998, Yu et al. 2007).

The metabolic assays described earlier have been based upon sensing molecules that were produced. It is possible to perform many of the assays described earlier on a molecular level by probing for mRNA of the desired transcripts.

Mechanical Activity

For many cell types, changes in dimension, position, or attachment characterize normal cell function. Assays that characterize this "mechanical activity" are commonly used post-thaw as a measure of function. The most common assay for mechanical activity is attachment. Cells are typically cryopreserved in suspension, and the ability of the cells to attach post-thaw is commonly performed. Anchorage-dependent cells incapable of attaching typically undergo a specialized programmed cell death known as anoikis.

Quantifying attachment of cells post-thaw can be fairly straightforward. A given number of cells are typically seeded using appropriate cell culture conditions for that cell type. After a given period of time (e.g., 2–4 h), unattached cells are removed with the supernatant and enumerated. The fraction of cells attached can then be determined.

The ability of certain cell types to move is also an important fundamental activity that denotes proper function. Motility assays are applied most commonly to sperm (De Jonge et al. 2003). For example, the characteristic movement (change in position or movement of tail) of at least 100 sperm can be documented and the percentage of sperm that are motile can be quantified in a sample.

Other types of motility assays include cell migration assays. The ability of cells to migrate is important in diverse applications including cancer (i.e., metastatic potential of cells), wound healing, and atherosclerosis. There are methods of quantifying migration. The two most common methods include a wound closure assay and a transwell migration assay to quantify invasion (Justus et al. 2014).

Proper function of muscle cells requires the ability to contract; therefore, contractility assays are commonly used for determining post-thaw function of muscle cells. Cardiomyocytes exhibit beating *in vitro* and this function is important for post-thaw recovery (Gu et al. 2013). Beating behavior can be quantified and analyzed using commercially available assays.

Another type of mechanical activity includes phagocytosis, used principally for characterizing function of macrophages and neutrophils. Phagocytosis is the process by which a cell engulfs a solid particle. Macrophages and other immune cells use this mechanism to remove pathogens or cell debris. Cells engulf a target particle (typically RBCs or zymosan particles) and the number of cells exhibiting this function are quantified using microscopy or targets that are not phagocytosed.

Aggregation (or self-assembly) of cells is also a mechanical assay. Hepatocytes, neuronal progenitor cells, embryoid bodies, beta-cells, ES, and iPS cells form either spherical or disk-shaped aggregates (Thomson et al. 1998, Yu et al. 2007). Coculture of fibroblasts and endothelial cells results in self-assembly into microcapillaries (Sorrell, Baber, and Caplan 2007). The ability of the cells to reassemble post-thaw into spheroids, disk-shaped colonies, or microcapillaries can be an assay for proper post-thaw mechanical function.

Mitotic Activity

The ability of a cell to proliferate post-thaw is another post-thaw metric that is commonly used. Determining proliferation post-thaw is very straightforward: a given number of cells are seeded into a culture environment after a given period of time (24–72h is common) and cells are enumerated after culture. An increase in cell number with time in culture demonstrates the ability of the cells to proliferate. It is possible to compare doubling time, T_d, for the cells pre-freeze and post-thaw to determine whether the cryopreservation process has altered the proliferative capability of the cells.

$$T_d = (t_1 - t_2) \times \log(2) / \log(N_2 / N_1)$$

where T_d is the doubling time, t_1 is the initial time, t_2 is the final time, N_2 is the number of cells at t_2, and N_1 is the number of cells at t_1.

Mitotic capability of cells can also be determined using flow cytometry (Anderson et al. 1998). For example, cells can be stained with a mitosis specific antigen, TG-3, and combined with quantitative DNA measurements or bromodeoxyuridine, which permits quantification of all four phases of the cell cycle (M, G_1, S, and G_2). Other techniques for determining mitotic activity using flow cytometry detect phosphorylated epitopes as biomarkers for mitotic cells (Jacobberger et al. 2008).

Differentiation Potential

By definition, a stem cell must exhibit the ability to differentiate into different cell types. Therefore, differentiation assays are commonly used post-thaw to verify that a stem cell is capable of differentiating post-thaw. Stem cells may be multipotent, pluripotent, or totipotent. Multipotent cells are capable of differentiating into more than one cell type. Adult stem cells, such as hematopoietic stem cells (HSCs), are capable of differentiating into all of the different blood cell types so they are considered multipotent. Pluripotent cells are capable of differentiating into the three germ layers (endoderm, mesoderm, and ectoderm) present in the body; ES cells are considered pluripotent. Totipotent cells are capable of forming all of the cell types in the body. Assays for totipotency that are widely accepted have not been developed.

Colony-forming unit (CFU) assays are commonly used to demonstrate the multipotency of hematopoietic progenitor cells (HPCs). Briefly, HPCs are seeded into a semiliquid methylcellulose medium supplemented with cytokines and then seeded into a culture plate. After a given culture period (10–14 days), colonies form and the colonies can be enumerated and scored for the presence of erythroid, granulocyte, granulocyte/macrophage, or megakaryocyte progenitor. This CFU assay is commonly performed on products that may be transplanted into patients receiving HSC transplants.

In contrast, mesenchymal stem cells (MSCs) are also multipotent adult stem cells, but they are expected to differentiate into fat, cartilage, and bone. The trilineage assay is used to demonstrate that capability. Briefly, MSCs are plated in three different containers and cultured with three different media, each designed to promote differentiation into adipocytes, chondrocytes, and osteocytes. After an appropriate culture period (7–21 days), the cultures are fixed and stained. Cultures for adipocytes are stained with Oil Red to determine the presence of lipid accumulation; chondrocyte cultures are stained with Alcian Blue to stain for sulfated glycosaminoglycans consistent with cartilage formation; and osteoblast cultures are stained with Alizarin Red to detect calcified extracellular matrix (Dominici et al. 2006). Trilineage assays are commonly performed on MSCs used clinically.

Other types of stem cells, such as ES and iPS cells, should be pluripotent and the teratoma assay is used to demonstrate that capability. ES or iPS cells are injected into the hind limb of a mouse and cultured for roughly 9 weeks (Thomson et al. 1998, Yu et al. 2007). Differentiation of the ES or iPS cells into the three different germ layers (ectoderm, mesoderm, and endoderm) is confirmed by histological staining of the injected region.

Transplantation Potential

The ultimate assay for a cell that is intended for therapeutic use is the ability of the cell to engraft *in vivo* and function as intended. HSCs are a common example of cells that are cryopreserved, infused, and the cells engraft and express mature blood cells. Over 30 000 people in the United States receive HSCs each year for the treatment of a variety of diseases (e.g., leukemia, Fanconis anemia, etc.) and many of those patients receive cryopreserved cells. The blood cells that can be found post-transplant contain the DNA of the donor, so it is known that the cells from the donor persist long-term.

It is common for transplantation assays to involve animal studies. Two different types of animal transplantation studies are commonly performed: syngenic and xenogenic transplantation studies. Syngenic studies involve transplantation of cells or tissues from an animal that is genetically identical or similar enough that there are no concerns regarding immunological response. Xenogeneic transplantation involves transplantation of cells from another species (typically human) into an animal model. Most xenogenic transplantation studies involve the use of immunocompromised animals.

Strategies to Improve the Accuracy and Reproducibility of Post-Thaw Assessment

Eliminate Measurement Bias

It is fairly common to insert a measurement bias into calculations of viability using common membrane integrity methods of assessment. This measurement bias is illustrated in Figure 7.1. In Figure 7.1a, 100 cells are present in the sample pre-freeze. A membrane integrity dye indicates that seven of the cells are not viable (cells that are grey in the figure). The overall viability of the sample is 93/100 or 93%. Post-thaw, the same assay is performed, and the total number of cells is 71 with five of the cells staining nonviable. Using the same method described earlier, the overall viability is calculated to be 66/71 = 93%. If the post-thaw viability is divided by the pre-freeze viability, the efficiency of the freeze-thaw process is calculated to be 100%.

Unfortunately, using this method of calculating viability or efficiency of your freeze–thaw process artificially elevates the outcome. For a population of "n" cells being cryopreserved, there will be three different populations:

1) Cells that are intact and viable.
2) Cells that are intact but *not* viable.
3) Cells that have lysed and are no longer intact.

(a)

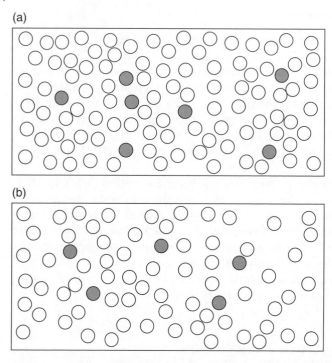

(b)

Figure 7.1 Pre-freeze and post-thaw assessment of a given sample. In this diagram, cells that are clear are viable and cells that are shaded grey are nonviable. The total number of cells in panel (a) is 100 and 93% of the cells are viable pre-freeze. The total number of cells in panel (b) is 71 and 93% of the cells are viable post-thaw. If cell losses are not included, the recovery of cells is artificially biased upward.

A common error when performing the calculations is to omit the cells that have lysed; therefore, they are not present in the sample. By omitting the population of cells that have lysed, the estimates for viability or efficiency of the freezing process have been artificially elevated. A more accurate measurement would be to estimate the recovery of the freezing process, R.

$$R = \frac{NPF}{NPT}$$

where NPF is the total number of viable cells pre-freeze and NPT is the total number of viable cells post-thaw.

Using the numbers given in the example previously, the recovery of the freezing process would be $66/93 = 71\%$ (versus 100% using the other method). *Therefore, calculating viability based solely on measurements of intact cells fails to account for cells that have lysed during the freezing process and results in an artificial upward bias in the viability calculated.*

- *Best practices for post-thaw assessment 1: Measure post-thaw recovery, which is defined as the number of viable cells post-thaw divided by the number of viable cells pre-freeze.* This practice eliminates the measurement bias associated with counting only intact cells.

Compensating for Post-Thaw Apoptosis

A decline in the number of cells as a function of time post-thaw has been documented for a wide variety of cell types (Figure 7.2). The decline in cell number results from post-thaw apoptosis (Baust et al. 2001, Stroh et al. 2002). The practical implications of this behavior is that timing of post-thaw assays, in particular physical integrity, must be consistent. Specifically, measuring post-thaw viability at 30 min post-thaw may result in a different outcome than measuring post-thaw assessment at 1 h post-thaw. Making the time for performing the protocol consistent removes that source of variability in the outcome of the protocol.

One option for quantifying cell number post-thaw may also involve waiting to assay viability until 24–36 h post-thaw. This time period permits the cells to die off and represents a conservative measure of post-thaw recovery.

- *Best practices for post-thaw assessment 2: Cell counts or assaying physical integrity using a membrane integrity dye should be performed at a consistent time post-thaw, and protocols should specify a narrow time window for performing the assay.* This practice will eliminate variation in the measurement resulting from post-thaw cell losses.

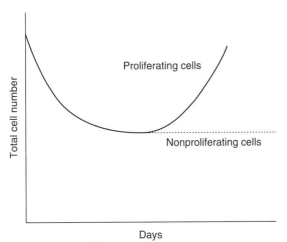

Figure 7.2 Variation in cell number with time post-thaw. Cells capable of proliferation will experience an increase in cell number after the rate of proliferation exceeds the rate of cell death. Cells that do not proliferate will initially die off and then cell numbers will plateau (dashed line).

Post-Thaw Assessment Using a Single Measure

During protocol development and optimization, another common error involves the use of a single measure for post-thaw assessment. The most common single measure used is physical integrity, most often, membrane integrity using dyes. This approach provides inadequate information for many different cell types. For example, hepatocytes can survive the freezing and thawing process with high levels of membrane integrity, but other functions including attachment and the ability to metabolize drugs are significantly diminished (Alexandre et al. 2012). As a result, the hepatocytes post-thaw do not function as needed. Therefore, using membrane integrity as a sole measure of post-thaw function does not reflect the actual efficiency of the preservation process since the cells may not exhibit the desired function post-thaw.

It is difficult and costly to perform a battery of post-thaw assays for every step of process development. A more practical approach is to perform a more in-depth battery of post-thaw assays on a regular basis to confirm the behavior observed with the rapid, easy-to-perform assay that is being used in development. Another approach is to use multiple measures if both are relatively easy to perform. For example, developing a method of preserving an anchorage-dependent cell could involve measuring both membrane integrity post-thaw and the ability of the cells to attach.

The use of a single measure to characterize post-thaw viability provides additional problems with heterogeneous cell populations. It is common to combine post-thaw assessment with a phenotypic characterization of the cells when heterogeneous cell populations are cryopreserved. It is important to determine if the desired subpopulations of cells survived the freezing protocol. HSC products are very heterogeneous and contain lymphocytes, granulocytes, monocytes, eosinophils, and a very small population of HPCs (1–2% $CD34^+CD45^+$ cells). The HPCs provide the therapeutic benefit (engraftment and expression of blood cells), and the remaining cells may support or suppress engraftment of the progenitor cells and maturation of blood cells. Therefore, post-thaw assessment may include phenotypic characterization to determine the presence of the target cells (i.e., HPCs) in the mixed population.

- *Best practice for post-thaw assessment* 3: Use of multiple measures of post-thaw assessment is critical, in particular during process development and optimization. At least one of those measures should reflect the desired post-thaw function of the cells.

Optical Methods of Post-Thaw Assessment

As described earlier in this chapter, cells will experience significant changes during the freezing process, which will influence the reliability of post-thaw assessment. In particular, optical methods of characterization (i.e., flow cytometry) may need recalibration or optimization specifically on frozen and thawed samples since freezing may influence the forward and side scattering as well as the complexity. These changes can influence the interpretation of the data.

- *Best practice for post-thaw assessment* 4: When using optical methods of post-thaw assessment, care should be used to recalibrate or alter data analysis to reflect the influence of freezing and thawing on the optical properties of cells.

Release Criteria

After thawing and post-thaw assessment, a decision must be made to either use or discard the sample. Each protocol should establish release criteria for downstream use of the cells. Standard release criteria for cell therapy products are a good starting point for establishing release criteria for any cryopreserved sample. In order to be acceptable for downstream use, a sample should meet established criteria for the following:

- Safety/sterility
- Identity
- Purity
- Potency

As described in Chapter 2, cells should be tested before freezing for bacteria, viruses, mycoplasma, or other adventitious agents. Additional safety testing may be performed post-thaw if there are concerns regarding potential contamination during the preservation process. Similarly, when the cells are tested before freezing for identity (Chapter 2), those studies can be used as part of the release criteria for the sample. The post-thaw assessment techniques described earlier most often address issues of purity and potency. These assays describe post-thaw viability and function. Acceptable levels of viability, cell recovery, or post-thaw function should be specified.

Scientific Principles

- Cells undergo profound changes during freezing, which may in turn influence the outcome and interpretation of post-thaw assessment.

Putting Principles into Practice

- Calculate cell recovery post-thaw in order to eliminate the measurement bias associated with only measuring viability of *intact* cells.
- Make sure that assays are performed at the same time post-thaw to eliminate variations that result from cells dying off post-thaw.

- Use multiple measures of post-thaw assessment, especially during protocol development and optimization. At least one of those measures should reflect the desired post-thaw function of the cell.
- When using optical methods of post-thaw assessment, calibration of the device or interpretation of the data should reflect changes in the optical properties of the cells resulting from the freezing process.
- Release criteria should be established appropriate for the downstream use of the cells for thawed cells.

References

Alexandre, E., A. Baze, C. Parmentier, C. Desbans, D. Pekthong, B. Gerin, C. Wack, P. Bachellier, B. Heyd, J. C. Weber, and L. Richert. 2012. "Plateable cryopreserved human hepatocytes for the assessment of cytochrome P450 inducibility: experimental condition-related variables affecting their response to inducers." *Xenobiotica* 42 (10):968–979.

Anderson, H. J., G. de Jong, I. Vincent, and M. Roberge. 1998. "Flow cytometry of mitotic cells." *Exp Cell Res* 238 (2):498–502.

Baust, J. M., M. J. Vogel, R. Van Buskirk, and J. G. Baust. 2001. "A molecular basis of cryopreservation failure and its modulation to improve cell survival." *Cell Transplant* 10 (7):561–571.

Chinnadurai, R., M. A. Garcia, Y. Sakurai, W. A. Lam, A. D. Kirk, J. Galipeau, and I. B. Copland. 2014. "Actin cytoskeletal disruption following cryopreservation alters the biodistribution of human mesenchymal stromal cells in vivo." *Stem Cell Rep* 3 (1):60–72.

De Jonge, C. J., G. M. Centola, M. L. Reed, R. B. Shabanowitz, S. D. Simon, and P. Quinn. 2003. "Human sperm survival assay as a bioassay for the assisted reproductive technologies laboratory." *J Androl* 24 (1):16–18.

Diepart, C., J. Verrax, P. B. Calderon, O. Feron, B. F. Jordan, and B. Gallez. 2010. "Comparison of methods for measuring oxygen consumption in tumor cells in vitro." *Anal Biochem* 396 (2):250–256.

Dominici, M., K. Le Blanc, I. Mueller, I. Slaper-Cortenbach, F. Marini, D. Krause, R. Deans, A. Keating, D. J. Prockop, and E. Horwitz. 2006. "Minimal criteria for defining multipotent mesenchymal stromal cells. The International Society for Cellular Therapy position statement." *Cytotherapy* 8 (4):315–317.

Dunn, J. C., M. L. Yarmush, H. G. Koebe, and R. G. Tompkins. 1989. "Hepatocyte function and extracellular matrix geometry: long-term culture in a sandwich configuration." *FASEB J* 3 (2):174–177.

Gu, Y., F. Yi, G. H. Liu, and J. C. Izpisua Belmonte. 2013. "Beating in a dish: new hopes for cardiomyocyte regeneration." *Cell Res* 23 (3):314–316.

Jacobberger, J. W., P. S. Frisa, R. M. Sramkoski, T. Stefan, K. E. Shults, and D. V. Soni. 2008. "A new biomarker for mitotic cells." *Cytometry A* 73 (1):5–15.

Justus, C. R., N. Leffler, M. Ruiz-Echevarria, and L. V. Yang. 2014. "In vitro cell migration and invasion assays." *J Vis Exp* e51046. doi: 10.3791/51046.

Pegg, D. E. 1989. "Viability assays for preserved cells, tissues, and organs." *Cryobiology* 26 (3):212–231.

Segnani, C., C. Ippolito, L. Antonioli, C. Pellegrini, C. Blandizzi, A. Dolfi, and N. Bernardini. 2015. "Histochemical detection of collagen fibers by Sirius Red/Fast Green is more sensitive than van Gieson or Sirius Red alone in normal and inflamed rat colon." *PLoS One* 10 (12):e0144630.

Sorrell, J. M., M. A. Baber, and A. I. Caplan. 2007. "A self-assembled fibroblast-endothelial cell co-culture system that supports in vitro vasculogenesis by both human umbilical vein endothelial cells and human dermal microvascular endothelial cells." *Cells Tissues Organs* 186 (3):157–168.

Stroh, C., U. Cassens, A. K. Samraj, W. Sibrowski, K. Schulze-Osthoff, and M. Los. 2002. "The role of caspases in cryoinjury: caspase inhibition strongly improves the recovery of cryopreserved hematopoietic and other cells." *FASEB J* 16 (12):1651–1653.

Thomson, J. A., J. Itskovitz-Eldor, S. S. Shapiro, M. A. Waknitz, J. J. Swiergiel, V. S. Marshall, and J. M. Jones. 1998. "Embryonic stem cell lines derived from human blastocysts." *Science* 282 (5391):1145–1147.

Yu, J., M. A. Vodyanik, K. Smuga-Otto, J. Antosiewicz-Bourget, J. L. Frane, S. Tian, J. Nie, G. A. Jonsdottir, V. Ruotti, R. Stewart, I. I. Slukvin, and J. A. Thomson. 2007. "Induced pluripotent stem cell lines derived from human somatic cells." *Science* 318 (5858):1917–1920.

8

Algorithm-Driven Protocol Optimization

As described in the previous chapters, survival of cells is strongly influenced by cooling rate. As described in Chapter 3, the post-thaw survival is influenced by both the solution composition and the cooling rate. The most common method of optimizing a protocol involves altering the solution composition and cooling rate, and then determining cell viability (Freimark et al. 2011). This methodology is expensive, time-consuming, and may not produce an optimal outcome.

Optimization of a process with independent variables (e.g., solution composition and cooling rate) and outputs (e.g., viability, post-thaw function, and cell surface phenotype) can be performed using a variety of methods. Differential evolution (DE) is a metaheuristic method of optimization that does *not* use a gradient implying that the approach can be used on problems that are not continuous or even change with time. DE requires minimal assumptions about the problem being solved and can operate with incomplete information.

Recently, a strategy 2 (DE/local-to-best/1, which balances robustness and convergence) DE algorithm was used to optimize a preservation protocol (Pollock et al. 2016b). This approach has been used to optimize the preservation of mesenchymal stromal cells without dimethlysulfoxide. This approach can be used to optimize both composition and cooling rate for a given cell type. Storn and Price (1997) developed the basic structure of the algorithm and it is available as open source software. This algorithm utilizes stochastic direct search and independently perturbs population vectors containing various solution compositions and cooling rates to identify a global maximum of post-thaw viability within the user-defined parameter space.

Generation 0 consists of a randomly generated initial population that spans the entire parameter space for the solution composition and cooling rate (Figure 8.1). This population is composed of a given number of independent parameters (solution composition and cooling rate) expressed as vectors.

Preservation of Cells: A Practical Manual, First Edition. Allison Hubel.
© 2018 John Wiley & Sons, Inc. Published 2018 by John Wiley & Sons, Inc.

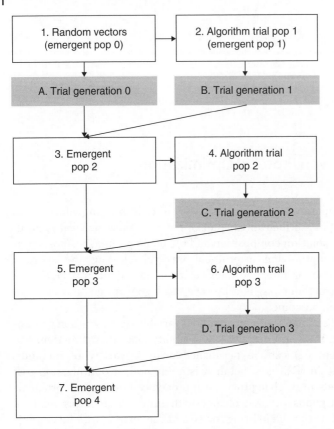

Figure 8.1 Flow chart of the DE algorithm. Black boxes represent DE algorithm steps and grey boxes represent experimental steps. The algorithm produces a vector in generation 0 that randomly spans the parameter space and a trial population generation 1 that is based on mutation of generation 0. Both of these vectors are tested experimentally and the corresponding live cell recovery is input into the DE algorithm, resulting in an emergent population (pop) and the process is repeated.

The size of the vector is determined by the number of different inputs that are being evaluated (e.g., the number of different solution components or cooling rates). Cells are then frozen using the compositions and cooling rates suggested by the generation 0 vector. The corresponding post-thaw recovery is measured.

The post-thaw recovery measured is modified using the algorithm and the resulting population represents a combination of composition and cooling rate that may result in more favorable cell recovery (i.e., new test vector). The best value from this comparison (either the original or the new mutant vector) is stored in an emergent population. This mutation/comparison process is repeated for all subsequent generations. The final emergent population contains a set of solutions that have all been independently optimized using stochastic direct search and includes the best possible independent parameters for freezing cells within the defined parameter space. Describing the mathematical basis for the DE approach and algorithm is beyond the scope of this chapter. Additional information on this method can be found in Storn, Price, and Lampinen (2005).

Convergence of the algorithm can be measured through two different metrics: (i) an increase in the cumulative best live cell recovery observed in a given population and (ii) a decrease in the number of improved solutions within the emergent population after each generation. Both of these parameters can be monitored as a function of generation during algorithm optimization.

The performance of the algorithm can be influenced by three different parameters: the generation size (NP), crossover rate (CR), and weighting (F). NP is, in this context, the number of solution compositions or cooling rates tested in a given vector. In general, NP must be larger than four. Increasing NP increases the probability of finding a global optimum but slows the rate of convergence. A rule of thumb is that NP can range between $3D$ and $8D$ (D is the number of independent variables being optimized).

CR controls which and how many components are mutated in each element of the current population. CR is a probability between 0 and 1. Convergence of the algorithm is accelerated with larger crossover rates. For multimodal, parameter-dependent problems, CR in the range of 0.9–1.0 is reasonable.

The weighting, F, is strictly greater than zero and it is common for $F > 0.4$. A larger value for F increases the probability of escaping a local optimum (to go on to find the global optimum). Typical values of F are between 0.4 and 0.95.

Small Cell Number/High Throughput Approach

After structuring the algorithm and determining NP, CR, and F, the next phase is to structure the freezing experiments. Cells in candidate solutions can be frozen in small volumes (approximately 100 µl) in 96-well plates. This approach reduces the number of cells and reagents involved in preservation. It is recommended that each experimental condition to be tested be repeated in triplicate, which enables identification of wells in which there are errors in terms of seeding or the concentration of cryoprotective agents seeded into the well.

Use of a 96-well plate requires characterization of the freezing and thawing behavior. Preliminary studies should be performed to characterize the spatial gradients during freezing and thawing across the plate using thermocouples. Wells with solutions to be tested should be limited to wells that have a fairly uniform freezing and thawing rate (i.e., wells on the periphery of the plate tend to cool and warm more rapidly than wells in the center). In order to prevent evaporation or contamination, the plates should be sealed using silicone well covers, which tolerates cryogenic temperatures.

The concentration of each component can vary discretely between 0 and the maxima that is identified from the literature or dictated by solubility or toxicity limits. The use of a lower value of zero would enable a component that did not contribute to cell recovery to be eliminated in the optimization.

This format for freezing also requires a high-throughput format for post-thaw assessment. The use of fluorescent dyes and a fluorescent plate reader enables rapid assessment of post-thaw viability for a large number of wells. Raw fluorescence values can be used to calculate the number of live and dead cells present in each well, by correlating the fluorescence measured to a control curve of unfrozen cells generated using serial dilution of known (live and dead) cell counts. The live cell recovery can be calculated by dividing the number of live cells present in thawed samples by the number of live cells seeded pre-freeze.

The approach of using membrane integrity as determined using fluorescent dyes is a quick method of determining post-thaw recovery. Alternative methods have also been used as described in detail in Chapter 7. Specifically, attachment of cells post-thaw has been used as a more functional method of post-thaw assessment (Pollock et al. 2017). Cells are thawed and then plated at known cell numbers into a culture plate. After an appropriate period of time for incubation, the plates are washed to remove any unattached cells and stained with a fluorescent dye and read with a plate reader. The percentage of attached cells can be determined as a function of the raw fluorescence when compared to a control curve of attached cells at a known density or cell number.

In general, rapid, easy methods of post-thaw assessment are used for optimization. Subsequent studies can be used to characterize post-thaw function in more detail using molecular or cellular assays that may be difficult to perform. If a particular function is not observed in the more detailed studies using an optimized solution, it may be beneficial to optimize the preservation using multiple metrics of post-thaw function.

Validating Operation of the Algorithm

Two different methods are commonly used to validate the algorithm. The first involves testing the entire range of composition and cooling rate and determining whether or not the optimum determined using the algorithm agrees with the optimum found when spanning the entire range of composition and cooling rate (Pollock et al. 2016b). The second method of validating the algorithm uses a different starting vector (generation 0) and determining whether the algorithm converges to the same value of the optimum from that different starting point. Keep in mind that the weighting (F) used may influence the ability of the solution to find the global maximum (versus a local maximum). Therefore, if changing the initial vector results in a different optimum after optimization of the algorithm, it may be helpful to evaluate the use of different weighting factors on the outcome as well.

Flexibility

The DE algorithm has been used to optimize composition and cooling rate. Studies to date have centered on the optimization of three-component solutions (Pollock et al. 2016a, Pollock et al. 2016b, Pollock et al. 2017), although five-component solutions have been optimized using DE.

Algorithm-driven optimization can be used to optimize a variety of other parameters in the protocol: duration of exposure to cryoprotective agents, cooling rate, nucleation temperature, and so forth. Increasing the number of independent variables will increase the generation size and the experimental complexity because experiments have to be constructed to test all of the variables across the parameter space.

The output metrics can also be expanded. Current approaches have centered on the post-thaw membrane integrity, recovery, or cell attachment. However, efficiency of a protocol can be optimized based on multiple post-thaw measures. For example, the preservation of hematopoietic progenitor cells could be optimized based on total cell recovery, percentage of cells expressing the $CD34^+CD45^+$ cell surface phenotype, and colony formation (conventional clinical measures for post-thaw recovery). Optimizing all of these metrics will mean that the protocol converges on the optimal combination of all post-thaw metrics, and not just the maximum of one single metric.

Practical Notes

The post-thaw recovery of cells frozen in 96-well plates can differ from those observed in vials and bags. As a result, validation of the protocol should be performed using the format (volume and cell number) for the actual downstream use.

It is possible for an optimum composition to include a "zero level" of a given cryoprotective agent. This outcome implies that this additive does not improve post-thaw recovery of the cells or may actually be harmful.

As post-thaw recovery cannot exceed 100%, it is possible to achieve a "best member" that results in an acceptable post-thaw recovery without convergence of the algorithm. Completion of the algorithm, however, can help map the range of optimal operating space. This knowledge could then be used to develop multiple protocols that could be used successfully.

Modeling in Cryobiology

A variety of processes have been modeled in the field of cryobiology, including transport of cryoprotective agents and water across the cell membrane, intracellular ice formation, and heat and mass transfer during freezing (see Benson

2015, for review). Other models have attempted to predict an optimum cooling rate for preservation based on a threshold value of intracellular supercooling and knowledge of the transport characteristics of a specific cell type (Kashuba, Benson, and Critser 2014, Woelders and Chaveiro 2004). The approach outlined in this chapter does not require *a priori* knowledge of the cell type or its biological or biophysical characteristics. These alternative approaches may be helpful when the DE approach outlined earlier does not result in an acceptable post-thaw recovery or if a specific cell type is not amenable to conventional cryopreservation protocols.

References

Benson, J. D. 2015. "Modeling and Optimization of Cryopreservation." In Cryopreservation and Freeze-Drying Protocols, edited by W.F. Wolkers and H. Oldenhof, 3rd ed., Methods in Molecular Biology 1257, 83–120. New York: Humana Press.

Freimark, D., C. Sehl, C. Weber, K. Hudel, P. Czermak, N. Hofmann, R. Spindler, and B. Glasmacher. 2011. "Systematic parameter optimization of a Me(2) SO- and serum-free cryopreservation protocol for human mesenchymal stem cells." *Cryobiology* 63 (2):67–75.

Kashuba, C. M., J. D. Benson, and J. K. Critser. 2014. "Rationally optimized cryopreservation of multiple mouse embryonic stem cell lines: II—Mathematical prediction and experimental validation of optimal cryopreservation protocols." *Cryobiology* 68 (2):176–184.

Pollock, K., Yu, G., Moller-Trane, R., Koran, M., Dosa, P. I., McKenna, D. H., and Hubel, A. (2016a) Combinations of Osmolytes, Including Monosaccharides, Disaccharides, and Sugar Alcohols Act in Concert During Cryopreservation to Improve Mesenchymal Stromal Cell Survival. *Tissue Eng Part C Methods* 22, 999–1008.

Pollock, K., J. W. Budenske, D. H. McKenna, P. I. Dosa, and A. Hubel. 2016b. "Algorithm-driven optimization of cryopreservation protocols for transfusion model cell types including Jurkat cells and mesenchymal stem cells." *J Tissue Eng Regen Med.* doi: 10.1002/term.2175.

Pollock, K., R. M. Samsonraj, A. Dudakovic, R. Thaler, A. Stumbras, D. H. McKenna, P. I. Dosa, A. J. van Wijnen, and A. Hubel. 2017. "Improved post-thaw function and epigenetic changes in mesenchymal stromal cells cryopreserved using multicomponent osmolyte solutions." *Stem Cells Dev* 26 (11):828–842.

Storn, R, and K. Price. 1997. "Differential evolution—a simple and efficient heurstic for global optimization over continuous spaces." *J Glob Optim* 11 (4):341–359.

Storn, R, K. Price, and J. A. Lampinen. 2005. Differential Evolution: A Practical Approach to Global Optimization. New York: Springer.

Woelders, H., and A. Chaveiro. 2004. "Theoretical prediction of 'optimal' freezing programmes." *Cryobiology* 49 (3):258–271.

Introduction

The following section contains several protocols that address cryopreservation, shipping, and thawing of cells. Each protocol contains certain common elements: equipment, supplies, cryoprotectant addition, freezing, and in certain situations, the thawing protocol. Several of the protocols discuss preparation of the cells prior to freezing or the expected quality of the cells prior to freezing. Several protocols include a section on safety measures to be used during preservation.

Different preservation technologies for freezing are described in the various protocols. Some of the protocols use controlled-rate freezing, others use passive freezing of samples. The oocyte vitrification protocol uses the cryoleaf vitrification system. Two different methods for the preservation of peripheral blood mononuclear cells are described. One method is more amenable for use with automated systems of liquid handling. Each system has advantages and limitations. Understanding the scientific principles behind a protocol will help the equipment selection process.

Different cell types require the use of different preservation protocols, and there may be more than one method of preserving a given cell type. For example, three different methods of preserving red blood cells are given. All three protocols result in an acceptable product. Therefore, cryopreservation protocols are fit-for-purpose. The steps and processes can vary with cell type, workflow, and resources.

The protocols may be useful for organizations wanting to preserve the cell type specified in the protocol. However, the real utility of the protocols is that they represent the real-world application of the scientific principles given earlier in the book. It will be important to understand the link between the chapters describing all the elements of the protocol and the actual steps of a preservation protocol. This understanding will enable users to (i) develop protocols and (ii) improve the performance of existing protocols.

Preservation of Cells: A Practical Manual, First Edition. Allison Hubel.
© 2018 John Wiley & Sons, Inc. Published 2018 by John Wiley & Sons, Inc.

Protocol Contributors

Thank you to the following individuals and groups that contributed protocols.

Cryopreservation of endothelial cells in suspension: Leah A. Marquez-Curtis, A. Billal Sultani, Locksley E. McGann, Janet A. W. Elliott, University of Alberta, Edmonton, Canada.

Cryopreservation of peripheral blood mononuclear cells from whole blood: Rohit Gupta and Holden Maeker, School of Medicine, Stanford University, Palo Alto, CA, USA.

Cryopreservation of human adipose stem cells: Melany Lopez and Ali Eroglu, Medical College of Georgia, Augusta, GA, USA.

Cryopreservation of red blood cells: Andreas Sputtek, Medical Laboratory Bremen, Bremen, Germany.

Cryopreservation of oocytes by slow freezing and oocyte vitrification and warming: Jeffrey Boldt, Community Health Network, Indianapolis, IN, USA.

Cryopreservation of hematopoietic progenitor cells, Thawing of hematopoietic progenitor cells, Transportation of hematopoietic progenitor cells and other cellular products, Thawing of hematopoietic progenitor cells, Cryopreservation of T-cells, and Thawing and reinfusion of T-cells were all provided by Jerome Ritz, Sara Nikiforow, and Mary Ann Kelley from Dana Farber Cancer Institute, Boston, MA, USA.

Cryopreservation of Endothelial Cells in Suspension

Principle

An optimized protocol for the cryopreservation of endothelial cells that has been validated for human umbilical vein endothelial cells and porcine and human corneal endothelial cells providing post-thaw viabilities above 85% (Marquez-Curtis et al. 2016, Marquez-Curtis, McGann, and Elliott 2017, Sultani et al. 2016). Endothelial cells form the inner lining of blood vessels and other tissues and provide a selectively permeable barrier. Vascular endothelial cells play pivotal roles in hemostasis, coagulation, immune responses, and angiogenesis and are often used to study dysfunctions and pathologies such as atherosclerosis and thrombosis.

Equipment and Supplies

Equipment

1) Analytical balance (Mettler-Toledo)
2) Biological safety cabinet (Western Scientific Services, Ltd)
3) Pipet-aid and Pipettors (Eppendorf)
4) Phase-contrast microscope (Labovert, Leitz)
5) Coulter counter or hemocytometer
6) Centrifuge (Eppendorf 5810R tabletop centrifuge, Hamburg, Germany)
7) Programmable methanol cooling bath (FTS Systems, Stone Ridge, NY, USA), or, alternatively, −80°C freezer (Revco Ultima II)
8) T-type thermocouple and OMB-DAQ-55 data acquisition module (OMEGA Engineering Inc, Stamford, CT, USA)
9) Water bath (VWR)
10) Timer
11) Liquid nitrogen (LN$_2$) Dewar

Preservation of Cells: A Practical Manual, First Edition. Allison Hubel.
© 2018 John Wiley & Sons, Inc. Published 2018 by John Wiley & Sons, Inc.

Supplies

1) 50-ml conical centrifuge tubes (Corning, Inc)
2) Pipettes and tips (200 μl, 1000 μl, Eppendorf)
3) Borosilicate glass culture tubes (6 × 50 mm, VWR)
4) O-rings and corks (VWR)
5) Styrofoam floaters
6) Plastic cryovials (Nunc, ThermoFisher Scientific, Inc)
7) Dimethylsulfoxide (DMSO; Fisher Scientific)
8) Hydroxyethyl starch (HES, Bristol-Myers Squibb), or, alternatively, Pentastarch, a 20% w/v HES solution (Preservation Solutions, Inc)
9) Endothelial cell media (LONZA)
10) Methanol (Fisher Scientific)
11) Liquid nitrogen (Praxair)
12) Long metal forceps
13) Styrofoam buckets to hold icewater and LN_2

Safety

1) Exercise caution when handling LN_2. Use only special containers designed for cryobiological use, typically double-walled pressure vessels. Always use stable four-wheeled carts when moving containers. Styrofoam boxes must be constructed with suitable wall thickness and loosely covered to allow for increased pressure. Exposure to the extremely low LN_2 temperature ($-196°C$) can cause frostbite and tissue damage, as well as asphyxiation because nitrogen is able to displace the oxygen in the air. Wear thick insulated loose-fitting gloves, goggles and full face shield, lab gown, full-length pants, and closed-toed shoes when dispensing LN_2 from a tank or a Dewar. Experiments using LN_2 should be carried out in a well-ventilated area. In case of a spill or leak of LN_2, isolate the area and stop the source of the leak if possible. Ventilate area or move exposed personnel to fresh air. In case of skin contact, immerse skin in circulating warm water ($37.8–43.3°C$) but do not use dry heat.
2) Methanol is flammable and should be kept away from open flames or other ignition sources. Keep stock in a designated flammable solvents storage cabinet. Keep the bath covered when not in use. Methanol is toxic if swallowed, absorbed through the skin, or inhaled. Basic personal protective equipment (gloves, safety goggles, lab gown, full-length pants, and closed-toed shoes) is required. Wash with copious amounts of water in case of skin contact and move the affected person to a ventilated area in case of inhalation.
3) DMSO is a combustible liquid, which may cause skin, eye, and respiratory tract irritation. It readily penetrates the skin and may carry other dissolved

materials into the body. Wear basic personal protective equipment; wash with copious amounts of water in case of skin contact.

4) Glassware can cause cuts and lacerations when it breaks. Inspect glassware for cracks before use. Always wear gloves and safety goggles. Loosen corks to relieve pressure build-up when transferring glass tubes from LN_2 to the 37°C water bath, and carry out the procedure at arm's length, away from the face.

5) Water baths can get contaminated with bacteria and/or fungi. Add anti-bacterial and antifungal drops during routine lab maintenance and always wear gloves.

6) Methanol bath, water bath, and thermocouple are electrical equipment, which may cause shock to the user. Exercise electrical safety precautions.

Procedure

Cell Preparation

Note: Cells can be isolated from tissues or purchased. If purchased, the cells are obtained cryopreserved, thawed, and cultured according to the manufacturer's instructions using the recommended media and other reagents. Cells isolated from tissue are also typically cultured prior to cryopreservation. Follow culture confluency and passage recommendations according to cell type.

1) Ensure that cells have reached appropriate confluency by examining the cultures under a phase-contrast microscope before trypsinization.

2) Cryopreserve cells when they are at an appropriate early passage and have the typical morphology of healthy growing cells. Overgrown and late-passage cells approach senescence and are not recommended for cryopreservation.

3) Take an aliquot of the cell suspension and perform a cell count using a Coulter counter/cell analyzer or a hemocytometer. Dilute the cells to a density of $1-2 \times 10^6$ cells/ml of media. If necessary, concentrate the cells by centrifugation at $200 \times g$ for 5 min at room temperature, remove the supernatant, add the appropriate amount of media, and resuspend the cells by gently pipetting up and down.

4) The cell suspensions may be kept on icewater, to prevent cell clumping, for 2–3 h after passaging if needed.

Preparation of Cryoprotectant Solution

Note: DMSO and HES are prepared at twice the concentration of the final solution as they are diluted two-fold upon mixing with an equal volume of the cell suspension. Because volume measurements are temperature-dependent, the amounts of components of the cryoprotectant solution are weighed.

Using Powdered HES

1) Weigh 5 ml of media in a 50-ml conical tube.
2) Add approximately 0.6 ml DMSO (approximately 0.66 g, assuming a density of 1.1 g/ml at room temperature) and record the weight.
3) Add approximately 0.75 g HES.
4) Calculate the weight percentages of DMSO (should be close to 10%) and HES (should be close to 12%).
5) Immerse the tube in a 37°C water bath and swirl occasionally to allow the HES to dissolve. Once complete dissolution is achieved, transfer the tube to icewater.

Using Pentastarch Solution

1) HES supplied by Bristol-Myers Squibb and used in our references 1 and 2 has been discontinued. As an alternative, we have validated Pentastarch (20% w/v, Preservation Solutions, Inc (PSI), Elkhorn, WI, USA). To prepare 10% DMSO and 12% Pentastarch in media, measure 2 ml media, 0.6 ml DMSO, and 3.75 ml PSI Pentastarch solution, taking note of the weight after each addition. Mix well by gently pipetting up and down. Calculate % by weight composition.

Cryoprotectant Addition

1) Weigh an equal volume of cell suspension and cryoprotectant solution. Calculate the weight percent of DMSO (should be close to 5%) and HES (should be close to 6%) in the final mixture.
2) Mix the cells and cryoprotectants by pipetting gently up and down. Leave the cell–cryoprotectant mixture in icewater for 15 min to allow the DMSO to penetrate the cells.

Freezing

Controlled-rate Freezing with a Methanol Bath

1) Set the programmable methanol cooling bath to −5°C with a mixing speed of about 70 rpm and a cooling rate of 1°C/min and begin the temperature data acquisition program to record the actual temperature in the methanol bath. Allow sufficient time for the bath to reach −5°C.
2) Aliquot 200 μl of the cell–cryoprotectant mixture to the borosilicate glass culture tubes (with O-rings to allow them to sit on a Styrofoam floater in the methanol bath) and cover with corks.
3) Place the culture tubes in the methanol bath and allow equilibration at −5°C for 2 min.

4) Induce ice nucleation by touching the tubes with LN_2–cooled metal forceps. Allow the release of the latent heat of fusion by keeping at −5°C for 3 min.

5) Set the methanol bath temperature to −50°C and allow controlled cooling at 1°C/min until about −35°C. (See the following for alternative procedure.)

6) Once the desired temperature is attained, transfer the tubes to LN_2.

Alternative Freezing Procedure

1) In case a programmable methanol–cooling bath or controlled-rate freezer is unavailable, we suggest as an alternative cooling 1 ml cell suspensions in the presence of 5% DMSO and 6% HES in plastic cryovials (Nunc 1.8 ml CryoTube vials, ThermoFisher Scientific, Waltham, MA, USA) placed in a polystyrene tube holder and kept in a −80°C freezer. We have validated the cooling rate to be 1.4 ± 0.3°C/min. The next day the cryovials can be transferred to LN_2 for long-term storage. After at least 8 months of storage we found that the post-thaw viability was similar to those of samples in glass tubes, cooled at 1°C/min in a controlled-rate methanol bath, that were plunged into LN_2. Commercially available cooling containers, such as Mr. Frosty (ThermoFisher Scientific) and CoolCell containers (Corning, Inc, Corning, NY, USA), in combination with a dry ice locker, can also provide a cooling rate of approximately 1°C/min.

Thawing

1) Rapidly thaw the cells in a 37°C water bath until only a sliver of ice remains.

Expected Results

Post-thaw membrane integrity for human umbilical vein endotheial cells (HUVECs) should be 87.7 ± 0.8%, assessed by flow cytometric analysis, which was equivalent to 94.0% when normalized against fresh, unfrozen control cells. The HUVECs should also exhibit tube-forming ability, an *in vitro* assay for angiogenesis. The post-thaw membrane integrity for porcine corneal endothelial cells should be near 85%.

References

Marquez-Curtis, L. A., McGann, L. E., and Elliott, J. A. W. (2017) Expansion and cryopreservation of porcine and human corneal endothelial cells. *Cryobiology.*

Marquez-Curtis LA, Sultani AB, McGann LE, Elliott JAW. Beyond membrane integrity: Assessing the functionality of human umbilical vein endothelial cells after cryopreservation. *Cryobiology.* 2016;72(3):183–90.

Sultani AB, Marquez-Curtis LA, Elliott JAW, McGann LE. Improved cryopreservation of human umbilical vein endothelial cells: A systematic approach. *Sci Rep.* 2016;6:34393.

Cryopreservation of Peripheral Blood Mononuclear Cells from Whole Blood

Principle

Peripheral blood mononuclear cells (PBMCs) are defined as any round nucleated cells present in our peripheral blood. In particular, these cell types include lymphocytes and monocytes, which play critical functions across multiple systems in humans. Basic life sciences have relied heavily on the use of PBMCs to help drive functional analysis in research for multiple decades.

The first methodology is the most common practice that can be found in the research community. The use of density centrifugation to separate the layers of whole blood and isolate PBMCs is typically accomplished by overlaying (or underlaying) a density gradient media, such as Ficoll-Hypaque. This is followed by a long centrifugation step, before carefully separating the PBMC layer, which, when done properly, will be clearly identifiable. Modern advancement in PBMC isolation has provided new mechanisms for processing whole blood. An additional method provided include the use of SepMates® (StemCell Technologies, Vancouver, British Columbia, Canada). This method allows for higher throughput and improved standardization than the traditional method.

PROTOCOL 1: ISOLATION OF PBMCS DIRECTLY OVER FICOLL-HYPAQUE

Equipment

- Benchtop centrifuge (Allegra X-15R, Beckman Coulter)
- Tali image-based cytometer (Invitrogen)
- Pipette gun (Drummond)
- p200 micropipette (Rainin)

Preservation of Cells: A Practical Manual, First Edition. Allison Hubel.
© 2018 John Wiley & Sons, Inc. Published 2018 by John Wiley & Sons, Inc.

Materials

- Heparin green top tube (Fisher, 367874)
- 1.8 ml cryotube vials (Fisher, 375418)
- 50 ml conical vial (Fisher, 352070)
- Sterile, filtered, p200 pipette tips (Rainin, RT L250F)
- CoolCell (Fisher, NC9883130) and CoolBox
- Tali cellular analysis slide (Invitrogen, 110794)
- 2, 5, 10, 25, and 50 ml sterile, serological pipettes (Fisher, 356507, 356543, 356551, 356525, 356550 respectively)
- Transfer pipette (Fisher, 357575)

Reagents

Ficoll-Paque PLUS (Fisher, 17-1440-03)
Ca+ and Mg+ Free PBS (Invitrogen, 10010-049)
Human serum type AB (Valley Biomedical, HP1022)
DMSO (Sigma-Aldrich, D8418-500ML)
Freezing media (refer to Appendix A)

Procedure

1) If plasma is needed prior to PBMC isolation, please refer to HIMC's Plasma Isolation SOP.
2) Pipette 15 ml of Ficoll into a new 50 ml conical tube.
3) Obtain whole blood from subject in heparin green top tubes.
4) Dilute whole blood 1:1 with PBS in a new 50 ml conical tube (NOTE: disregard this step if plasma was already isolated from Step 1).
5) Add heparinized whole blood to the conical tube by slowly pipetting it down the side of the tube, layering on top of the Ficoll.
   a) Add no more than 35 ml of diluted blood to the tube.
      i) If necessary, split the sample into two conical tubes.
6) Centrifuge the tubes at $800 \times g$ for 20 min with the brake off.
7) Remove tubes carefully from centrifuge.
8) Use a transfer pipette and remove the buffy coat into a new 50 ml conical vial. Take caution not to draw up the layers below the buffy coat.
   a) If a granulocyte pellet is needed, do not throw conical with ficolled blood away. Refer to HIMC's Granulocyte Isolation SOP.
9) Add PBS to the tube up to the 50 ml mark.
10) Centrifuge the PBMCs at $250 \times g$ for 10 min.
11) Aspirate the supernatant and resuspend the cells in 48 ml of PBS.

12) Count the cells using the Tali Counter (or lab's preferred cell counting method).
    a) Add 25 μl of the cell suspension to a Tali slide.
    b) Choose the "Quick Count" selection and "Name Now."
    c) Label the data with the sample ID.
    d) Insert slide into the Tali following the arrows on the slide.
    e) Press the button "Press to Insert New Sample."
    f) Focus the image so that the cells can be seen clearly with definitive borders.
    g) Press "Press to Run Sample."
    h) After counting, set the cell size to "5 μm to 15 μm" (this only has to be done to the first sample of the day).
    i) Calculate the total cell count by multiplying the number of cells/ml by the total volume of cell suspension. For example, 3.45×10^5 cells/ml $\times$ 48 ml = 165.6×10^5 cells.
13) Centrifuge the conical vial at $250 \times g$ for 10 min.
    a) Based on the total cell count, calculate the number of vials and volume of freezing media that will be needed.
        i) Label the appropriate number of empty cryovials with de-identified cryogenic label and place in a CoolBox to chill for at least 10 min (alternatively 4°C/wet ice can be used)
        ii) Pull enough Freezing Media A and Freezing Media B to create 1 ml aliquots. The total amount of freezing media needed is equal to the total number of aliquots needed.
14) Aspirate the supernatant.
15) Resuspend the cells in Freezing Media A equal to one half of the total freezing media needed.
16) Using a dropwise technique (1 drop/s) while swirling the sample, add Freezing Media B equal to the remaining half of the total volume.
17) Aliquot 1 ml of cell suspension into each cryovial.
18) Place the cryovials into a CoolCell and into a −80° freezer for 24 h (alternatively, a Mr. Frosty or controlled-rate freezer can be used).
19) Following this, immediately put the PBMCs cryovials into liquid nitrogen (LN₂) for long-term storage.

PROTOCOL 2: ISOLATION OF PBMCS USING SEPMATES

Equipment

- Benchtop centrifuge (Allegra X-15R, Beckman Coulter)
- Tali image-based cytometer (Invitrogen)
- Pipette gun (Drummond)
- p200 micropipette (Rainin)

Materials

- Heparin green top tube (Fisher, 367874)
- 1.8 ml cryotube vials (Fisher, 375418)
- 50 ml conical tube (Fisher, 352070)
- 50 ml SepMate tube (StemCell, 15450)
- 15 ml SepMate tube (StemCell, 15415)
- Sterile, filtered, p200 pipette tips (Rainin, RT L250F)
- Tali cellular analysis slide (Invitrogen, 110794)
- CoolCell (Fisher, NC9883130) and CoolBox
- 2, 5, 10, 25, and 50 ml sterile, serological pipettes (Fisher, 356507, 356543, 356551, 356525, 356550 respectively)

Reagents

- Ficoll-Paque PLUS (Fisher, 17-1440-03)
- Ca^+ and Mg^+ Free PBS (Invitrogen, 10010-049)
- Human serum type AB (Valley Biomedical, HP1022)
- DMSO (Sigma-Aldrich, D8418-500ML)
- Freezing media (refer to Appendix A)

Procedure

1) If plasma is needed prior to PBMC isolation, please refer to HIMC's Plasma Isolation SOP.
2) Pipette 15 ml of Ficoll into the central hole of a 50 ml SepMate tube.
   a) If there is 5 ml or less of whole blood, use a 15 ml SepMate tube and 4.5 ml of Ficoll.
3) In a 50 ml conical tube, measure the volume of heparinized whole blood and add an equal volume of PBS (Note: disregard this step if plasma was already isolated from Step 1).
4) Add the blood to the SepMate tube by pipetting it down the side of the tube
   a) Add no more than 34 ml of blood (no more than 17 ml whole blood)
   b) Add no more than 10 ml of blood (no more than 5 ml whole blood) with a 15 ml SepMate.
5) Centrifuge the vial at 1200×g for 10 min with the brake on.
6) Invert the tube (for no longer than 2 s) and pour the plasma and PBMCs into a new 50 ml conical vial.
7) Add PBS to the tube up to the 50 ml mark.
8) Centrifuge the PBMCs at 250×g for 10 min.
9) Aspirate the supernatant and resuspend the cells in 48 ml of PBS.

10) Count the cells using the Tali Counter (or lab's preferred cell counting method).
    a) Add 25 µl of the cell suspension to a Tali slide.
    b) Choose the "Quick Count" selection and "Name Now."
    c) Label the data with the sample ID.
    d) Insert slide into the Tali following the arrows on the slide.
    e) Press the button "Press to Insert New Sample."
    f) Focus the image so that the cells can be seen clearly with definitive borders.
    g) Press "Press to Run Sample."
    h) After counting, set the cell size to "5 µm to 15 µm" (this only has to be done to the first sample of the day).
    i) Calculate the total cell count by multiplying the number of cells/ml by the total volume of cell suspension. For example, 3.45×10^5 cells/ml $\times$ 48 ml $= 165.6 \times 10^5$ cells.
11) Centrifuge the conical vial at $250 \times g$ for 10 min.
    a) Based on the total cell count, calculate the number of vials and volume of freezing media that will be needed.
        i) Label the appropriate number of empty cryovials with de-identified cryogenic label and place in a CoolBox to chill for at least 10 min (alternatively, 4 C/wet ice can be used).
        ii) Pull enough Freezing Media A and Freezing Media B to create 1 ml aliquots. The total amount of freezing media needed is equal to the total number of aliquots needed.
12) Aspirate the supernatant.
13) Resuspend the cells in Freezing Media A equal to one half of the total freezing media needed.
14) Using a dropwise technique (1 drop/s) while swirling the sample, add Freezing Media B equal to the remaining half of the total volume.
15) Aliquot 1 ml of cell suspension into each cryovial.
16) Place the cryovials into a CoolCell and into a −80° freezer for 24 h (alternatively, a Mr. Frosty or controlled-rate freezer can be used).
17) Following this, immediately put the PBMCs cryovials into LN_2 for long-term storage.

APPENDIX A   Human Serum AB Freezing Media

Materials

- 50 ml conical vial (Fisher, 352070)
- 0.2 µm filter unit (Fisher, SCGPU02RE)
- 15 ml conical vial (Fisher, 1495949B)
- 2, 5, 10, 25, and 50 ml sterile, serological pipettes (Fisher, 356507, 356543, 356551, 356525, 356550 respectively)

Equipment

- Pipette gun (Drummond)
- Water bath

Reagents

- Human serum type AB (Valley Biomedical, HP1022)
- DMSO (Sigma-Aldrich, D8418-500ML)

Procedure

1) Thaw one bottle of Human serum AB.
2) Set the water bath temperature to 56°C.
3) Attach filter unit to the vacuum line in the biological safety cabinet, pour serum into filter unit, and filter through until all the media is filtered.
   a) If unit becomes clogged, a second filter unit may be necessary.
4) In 50 ml conicals, make two 30 ml aliquots of serum and two aliquots of 20 ml.
5) Place tubes in 56°C water bath and heat. Inactivate them for 30 min, swirling the tubes every 5–10 min.
6) While the tubes are in the water bath, prepare 15 ml conical vials, labeling them "A" or "B."
7) Remove the serum from the water bath and allow the conicals containing 20 ml to cool.
8) **Freezing Media A**—100% Human serum AB
   a) Aliquot the serum from the 30 ml tubes into the 15 ml conicals, 5 ml per tube—approximately a total of 12 tubes.
9) **Freezing Media B**—80% Human serum AB + 20% DMSO
   a) Add 5 ml of DMSO to the cooled serum.
      i) Add it dropwise while swirling to prevent precipitation.
      ii) Capping and inverting several times will help prevent precipitates.
   b) Aliquot the serum into 15 ml conicals, 5 ml per tube—approximately a total of 10 tubes.
10) Place all aliquots into a –20°C freezer until use. Thaw for use in cryopreservation and do not refreeze.

Cryopreservation of Human Adipose Stem Cells

Principle

Adipose-derived stem cells (ASCs) reside in the stromal compartment of adipose tissue and can be easily harvested in large quantities through a clinically safe liposuction procedure. ASCs hold great promise for cell-based therapies and tissue engineering. The protocol described below is a defined and xeno-free cryopreservation methodology (Lopez et al. 2016).

Equipment and Supplies

To carry out the complete procedure, the following items are required in addition to standard laboratory equipment and tools:

- Biological safety cabinet
- Controlled-rate freezer
- −80°C Mechanical freezer
- NALGENE® Mr. Frosty freezing container (cat. no. 5100-0036) or similar 1°C/min-freezing containers
- Autoclave
- Eye protection
- Cryogloves (face mask or goggles)
- Double impulse heat sealer (American International Electric, Inc)
- Surgery tools (e.g., scissors, scalpels, forceps, etc.)
- Kimwipes
- Liquid nitrogen (LN$_2$)
- Liquid nitrogen Dewar
- Conical centrifuge tubes, 50 ml, sterile (Nunc, cat. no. 339652 or similar)
- Conical centrifuge tubes, 20 ml, sterile (Nunc, cat. no. 339650 or similar)
- Cryovials (Nalgene, cat. no. 5000-0020 or similar)
- Cryostraws, 1/2 cc (TS Scientific, cat. no. TS202 or similar)

Preservation of Cells: A Practical Manual, First Edition. Allison Hubel.
© 2018 John Wiley & Sons, Inc. Published 2018 by John Wiley & Sons, Inc.

- Falcon 40-μm cell strainers (Becton Dickinson, cat. no. 352360 or similar)
- MidiMACS separator (cat. no. 130-042-302, Miltenyi Biotec)
- MACS LD columns (cat. no. 130-042-901, Miltenyi Biotec)
- MACS anti-FITC Microbeads (cat. no. 130-048-701, Miltenyi Biotec)

Reagents and Media

- Hanks Balanced Salt Solution containing 1.26 mM calcium and 0.90 mM magnesium without phenol red (HBSS, Gibco, cat. no. 14025-050 or similar)
- Dulbecco's calcium- and magnesium-free phosphate-buffered saline (DPBS-, Gibco, cat. no. 14190-144)
- Versene (Gibco, cat. no. 15040-066 or similar)
- Recombinant trypsin EDTA solution (Biological Industries, cat. no. 03-079-1C or similar)
- Defined trypsin inhibitor (Sigma, cat. no. T7659 or similar)
- Collagenase A type I (Sigma-Aldrich, cat. no. C-0130)[1]
- Ficoll-Paque Premium 1.073 (GE Healthcare, cat. no. 17-5442-52 or similar)
- 100 × Antibiotic-Antimycotic Mix (Gibco, cat. no. 15240062 or similar)
- 100 × Glutamax (Gibco, cat. no. 35050-061 or similar)
- 100 × Non-essential amino acids (NEAA, Gibco, cat. no. 11140-050 or similar)
- 50 × Essential amino acids (EAA, Gibco, cat. no. 11130-051 or similar)
- Ethylene diamine tetraaceticacid tetrasodium salt (EDTA, Sigma, cat. no. E6511 or similar)
- Ethylene glycol tetraaceticacid (EGTA, Fluka, cat. no. 03778 or similar)
- Dimethylsulfoxide (DMSO, Sigma, cat. no. D8418 or similar)
- Ethylene glycol (EG, Fluka, cat. no. 03750 or similar)
- Trehalose (Sigma, cat. no. T9531 or similar)
- Ficoll (Sigma, cat. no. F2878 or similar)
- Polyvinyl alcohol (PVA, Sigma, cat. no. P8136 or similar)
- Glutathione (Sigma, cat. no. G4251 or similar)
- L-Ascorbic acid 2-phosphate magnesium salt (Wako, cat. no. 013-12061 or similar)
- 70% ethanol
- CD45-FITC, anti-human monoclonal antibody (cat. no. 130-080-202, Miltenyi Biotec)
- CD31 FITC, recombinant human antibody (cat. no. 130-110-806, Miltenyi Biotec)
- *Column buffer*: DPBS (pH 7.2) containing 0.5% PVA and 2 mM EDTA.
- Leibovitz's L-15 Medium (Gibco, cat. no. 11415-064 or similar)
- CTS Knockout Dulbecco's modified Eagle's medium/Ham's F-12 mixture (DMEM/F-12, Gibco, cat. no. A13708-01 or similar)
- DMEM/F-12, HEPES (Gibco, cat. no. 11330032 or similar)

- *50% (w/v) collagenase stock solution (0.5 g/ml)*: Weigh out 1 g of type I collagenase and dissolve it in 2 ml of HBSS containing calcium and magnesium. Sterile filter using a 0.2-μm syringe filter, aliquot into sterile microcentrifuge tubes, and store at −20°C or below.
- *1000 × (0.1 M) EDTA stock solution*: Dissolve 0.42 g EDTA tetra sodium salt in 10 ml DPBS, sterile filter using a 0.2-μm syringe filter, and store at 4°C.
- *Red blood lysis buffer*: Prepare by dissolving 155 mM NH_4Cl (Sigma), 10 mM $KHCO_3$ (Sigma), and 1 mM EDTA in ultrapure water (pH 7.3) and sterilize by filtering through a 2-μm polyethersulfone membrane filter.
- *HBSS containing 0.03% PVA*: Dissolve 0.03 g PVA in HBSS containing calcium and magnesium (Gibco, cat. no. 14025-050), sterile filter using a 0.2-μm syringe filter, and store at 4°C.
- *Xeno-free defined cryopreservation medium without penetrating cryoprotectants*: Add 3 mM reduced glutathione, 5 mM ascorbic acid 2-phosphate, 0.25 M trehalose, 2% PVA, 5% ficoll, and 0.1 mM EGTA to HEPES-buffered DMEM/F-12 and filter under the hood using 0.2 μm membrane filter).
- *Xeno-free defined cryopreservation medium containing 2 × penetrating cryoprotectants*[2]: Add 10% DMSO, 10% EG, 3 mM reduced glutathione, 5 mM ascorbic acid 2-phosphate, 0.25 M trehalose, 2% PVA, 5% ficoll, and 0.1 mM EGTA to HEPES-buffered DMEM/F-12 and filter under the hood using 0.2 μm nylon membrane filter).

Procedure

Isolation of Human ASCs from Lipoaspirate[3]

1) With the proper personal protective equipment on, take specimen container and spray down with ethanol. Place it inside biosafety hood.
2) Transfer an appropriate volume of lipoaspirate to a new sterile container and wash with an equal volume of warm HBSS containing 1 × antibiotic–antimycotic mix three to four times to remove excess blood cells. Rigorously shake the container each time and then allow phase separation for 3–5 min. The adipose tissue will float above the HBSS now containing blood cells. Carefully aspirate the HBSS (blood cells) with a 50 ml pipette and repeat the wash until the HBSS from the final wash is clear.
3) Prepare a diluted collagenase solution from the 50% stock solution prior to the enzymatic digestion step (final working concentration of collagenase is 0.1%). Typically, the volume of the enzyme solution required is half that of the lipoaspirate volume. For example, if the lipoaspirate volume is 20 ml, dilute 60 μl of the collagenase stock solution in 10 ml HBSS containing typical concentrations of calcium and magnesium and 0.03% PVA. Next, add the

diluted collagenase solution to the washed lipoaspirate to have 0.1% final collagenase concentration and mix it by shaking the bottle vigorously.

4) Place the lipoaspirate container in a 37°C shaking water bath at approximately 75 rpm for 40–60 min until the fat tissue appears smooth on visual inspection.

5) During the collagenase treatment, prepare Ficoll-Paque gradients by dispensing 4 ml of Ficoll-Paque Premium 1.073 into 15 ml tubes.[4] The gradients must be equilibrated at room temperature (RT) before use.

6) After digestion, transfer the lipoaspirate container to the biosafety hood and add 0.1 M EDTA stock solution to a final concentration of 0.1 µM to stop collagenase activity. Next, filter the digested lipoaspirate through a 1 mm sterile sieve into 50 ml tubes.[5]

7) Spin samples at $300 \times g$ in an appropriate centrifuge for 5 min at RT. Thereafter, shake samples vigorously to disrupt the pellet. This is done in an effort to complete the separation of the stromal cells from the primary adipocytes.

8) Repeat centrifugation and carefully remove top layer of oil and fat, primary adipocytes (which will appear as a yellow layer of floating cells), and the underlying layer of collagenase solution. Leave behind small volume of solution above the pellet, so that the cells of the stromal vascular fraction (SVF) are not disturbed. The cell pellet usually includes a layer of dark red blood cells and appears as a red/pink color.

9) Resuspend the cell pellet in 20 ml of HBSS containing antibiotic–antimycotic mix and centrifuge at $300 \times g$ for 5 min.

10) Aspirate supernatant without disturbing the cell pellet.[6]

11) To eliminate red blood cells, resuspend the cell pellet in 20 ml of red blood cell lysis buffer. Incubate at RT for 10 min.

12) Centrifuge at $300 \times g$ for 10 min and aspirate cell lysis buffer.

13) Resuspend SVF in HBSS and pass through a 40-µm cell strainer.

14) Adjust the volume of cell suspension such that each ficoll-paque gradient tube receives 9 ml cell suspension (total volume per gradient: 13 ml).

15) Hold a gradient tube containing 4 ml ficoll-paque at a 45° angle and slowly add 9 ml cell suspension to form a layer without any mixing. Proper layering is critical for successful cell separation.

16) Centrifuge at $400 \times g$ for 30 min.

17) Aspirate the top HBSS fraction (approximately 8 ml) above the white band of cells found at the gradient interface and discard.

18) Carefully aspirate the white band of cells (2–5 ml) and release into a 50-ml tube containing 25 ml HBSS.

19) Centrifuge at $300 \times g$ for 10 min.

20) Repeat the washing step by resuspending the cell pellet in fresh HBSS to eliminate ficoll-paque taken while aspirating the white band of cells at the interphase.

21) Centrifuge at $300 \times g$ for 10 min, resuspend the cell pellet in HBSS, and determine cell count and viability using Trypan Blue staining or similar approaches. After this step, the isolated cells can be first subjected to sorting (optional), and then be cryopreserved or expanded *in vitro*.

Magnetic Cell Sorting (Optional)

To further enrich ASCs, the cells collected by density gradient separation can be subjected to magnetic sorting and negative selection. To this end, unwanted endothelial cells ($CD31^+$) and leukocytes ($CD45^+$) can be labeled first with FITC-conjugated anti-CD31 and anti-CD45 antibodies, and then with MACS anti-FITC microbeads. When applied to MACS LD columns placed on a Midi MACS separator, the magnetically labeled $CD31^+$ (endothelial cells) and $CD45^+$ (leukocytes) are retained in the columns, while unlabeled $CD31^-CD45^-$ ASCs are passed through and collected in a tube for subsequent use.

1) Wipe down the Midi MACS separator with 70% EtOH and introduce under the hood.
2) Resuspend approximately 10^7 cells collected by density gradient separation in 100 μl of column buffer and label with anti CD31-FITC and anti CD45-FITC antibodies by adding 10 μl of each antibody and incubating at 4°C for 15 min.
3) Aspirate unbound antibodies after adding 2 ml of column buffer and subsequent centrifugation at $300 \times g$ for 10 min.
4) Resuspend the cell pellet in 90 μl of column buffer, add 10 μl of MACS anti-FITC magnetic microbeads and incubate at 4°C for 15 min.
5) Aspirate unbound microbeads after adding 2 ml of column buffer and subsequent centrifugation at $300 \times g$ for 10 min at 4°C.
6) Resuspend the cell pellet in 500 μl of column buffer and place a MACS LD column on MidiMACS separator for each cell pellet.
7) Prepare each column by washing with 2 ml column buffer.
8) Pipette each cell suspension into individual columns and collect unlabeled cells flowing through the columns into 15-ml tubes.
9) Add 1-ml column buffer to each column to wash out all unlabeled cells.

Cryopreservation

Controlled-rate Freezing of Human ASCs[7]

1) Prepare cryopreservation media in advance. When protected from light and stored at 4°C, cryopreservation media are good for 2 weeks. Therefore, preparation of cryopreservation media in small volumes is recommended.
2) Enter the following ASC cooling program into the controlled-rate freezer:
   Step 1: Rapid cooling (10°C/min or 20°C/min) from RT to the start temperature (i.e., 0°C).

Step 2: Hold at 0°C until samples are introduced and the program is activated.

Step 3: Cooling to −7°C at 2°C/min.

Step 4: Soak time for 2 min and subsequent manual seeding of extracellular ice at −7°C.

Step 5: Holding at −7°C for 10 min.

Step 6: Cooling to −70°C at 1°C/min and holding at −70°C until transferring samples to LN_2.

3) Label cryovials or irradiated sterile 0.5-cc straws under the hood.[8]

4) Open LN_2 supply valve, turn on the controlled-rate freezer and run the ASC cryopreservation program to precool the chamber to the start temperature.

5) When *in vitro* expanding, trypsinize ASCs upon reaching 80% confluency. To do so, first replace culture medium with versene. Next, aspirate versene and add an appropriate volume of defined trypsin-EDTA solution at RT. Observe detachment of ASCs under microscope. Do not wait until complete detachment of ASCs. Facilitate detachment by tapping the culture dish. After approximately 10 min, add an equivalent amount of soybean trypsin inhibitor to stop trypsin activity. Complete cell detachment by aspirating and forcefully releasing cell suspension on culture surface. Finally, transfer cell suspension in a centrifuge tube and pellet the cells at $300 \times g$ for 10 min.

6) Discard supernatant and resuspend the ASC pellet in xeno-free defined cryopreservation medium containing no penetrating cryoprotectants. Perform cell count and centrifuge at $300 \times g$ for 10 min.

7) Discard supernatant and resuspend the ASC pellet in fresh xeno-free defined cryopreservation medium containing no penetrating cryoprotectants.

8) Add drop-wise an equal volume of the xeno-free defined cryopreservation medium containing $2 \times$ penetrating cryoprotectants while swaying the tube to gently mix the cell suspension (final penetrating cryoprotectant concentrations: 5% DMSO and 5% EG).

9) Allow the cells to equilibrate with $1 \times$ cryopreservation medium at RT for 10 min and load the cells into 0.5-cc straws or cryovials during the equilibration period as instructed next.

10) Gently mix the cell suspension and aseptically load 0.5-cc straws under the hood in the following order:

- Connect a 1-ml syringe to the cotton-plugged end of a 0.5-cc straw via short silicon tubing.
- Aspirate 1 cm column of $1 \times$ cryopreservation medium.
- Aspirate 1 cm air column.
- Aspirate 7–8 cm cell suspension column.
- Finally, aspirate 2–3 cm air column. This should result in wetting white powder at the cotton-plugged end with the first 1-cm column of the cryopreservation medium.
- Seal both ends of the straw with Double Impulse Sealer.

When using cryovials, mix and dispense 1-ml cell suspension in each cryovial.

11) At the end of the equilibration time, place two straws (or two cryvials) in each sample holder, introduce the sample holders into controlled-rate freezer and reactivate the freezing program to cool the samples from 0°C to the seeding temperature (i.e., −7°C). While the program is running, fill a styrofoam cup with LN_2 to prechill forceps for seeding.[9]

12) Upon cooling to −7°C and following the soak time, seed extracellular ice by touching the prechilled forceps to the powder end of each straw (or to the wall of each cryovial).

13) Reactivate the cooling program to complete the subsequent steps.

14) Upon cooling to −70°C, remove and plunge sample holders into LN_2 without raising sample temperature.

15) End the cooling program and allow controlled-rate freezer to return RT before turning off.

Thawing Human ASCs[10]

1) To thaw ASCs frozen in 0.5-cc straws:
   - Remove a straw from LN_2 and thaw its content at RT for approximately 3 min by horizontally placing it on a kimwipe under the hood.
   - Wipe down the outside of the straw with 70% ethanol and aseptically cut off both ends.
   - Release cells into a 15-ml tube by pushing on cotton end with a long blunt needle.

2) To thaw ASCs frozen in cryovials:
   - Remove a cryovial from LN_2 and hold in air for 20 s. Then, partially immerse it in a water bath at 37°C and swirl until ice melts.
   - Wipe down the outside of the cryovial with 70% ethanol.
   - Gently mix cell suspension under the hood and transfer to a 15-ml centrifuge tube.

3) To dilute cryoprotectants, add an equal amount of defined cryopreservation medium without penetrating cryoprotectants (1:1 dilution), gently swirl the tube, and hold at RT for 5 min.

4) Repeat the 1:1 dilution step by adding an equal amount of defined cryopreservation medium without penetrating cryoprotectants, gently swirl the tube, and hold at RT for 5 min.

5) For the final dilution step, add excess amount (e.g., 8–10 ml) of plain HEPES-buffered DMEM/F-12, gently mix, and wait for 5 min.

6) Pellet ASCs by centrifugation at $300 \times g$ for 10 min.

7) Aspirate supernatant, resuspend ASCs in appropriate medium, and determine cell number and viability.

Notes

1 GMP-grade collagenase is available for clinical applications (e.g., Serva collagenase NB 6, cat. no. 17458).

2 To reduce osmotic stresses, it is recommended to first resuspend a cell pellet in xeno-free defined cryopreservation medium without penetrating cryoprotectants and then drop-wise add equal volume of xeno-free defined cryopreservation medium containing $2\times$ penetrating cryoprotectants to have final xeno-free defined cryopreservation medium consisting of 5% DMSO, 5% EG, 3 mM reduced glutathione, 5 mM ascorbic acid 2-phosphate, 0.25 M trehalose, 2% PVA, 5% ficoll, and 0.1 mM EGTA in HEPES-buffered DMEM/F-12.

3 Working with human tissues and cells: all work with human cells and tissues should be done according to Biosafety Level 2 (BSL-2) practices and containment. It is necessary to obtain approval from the institutional biosafety committee and develop standard operating procedures based on guidelines for BSL-2. Extreme care must be taken to avoid aerosol-producing procedures, spilling, and splashing when working with any of these materials. Pathogens should be presumed in/on all equipment and devices that come into direct contact with any of these materials. All human material should be decontaminated by autoclaving or disinfection before discarding.

4 To preferentially isolate mesenchymal stem cells, a density gradient separation using ficoll-paque premium 1.073 has been suggested although other formulations also give satisfactory results.

5 To avoid spilling of the digested lipoaspirate, the metal sieve with 1-mm pores could be placed on top of a sterile funnel that in turn could be placed on top of a 50-ml tube.

6 When aspirating, the tip of the pipette should aspirate from the top, so that the oil is removed as thoroughly as possible.

7 Cooling and warming rates can be better controlled when using a controlled-rate freezer. Importantly, a controlled-rate freezer allows deliberate seeding of extracellular ice at a programmed temperature close to the freezing point of a given cryopreservation medium. Although not systematically compared, we also obtained satisfactory viability rates after freezing of human ASCs in Mr. Frosty freezing container that provides cooling rates around 1°C/min upon placing in a −80°C freezer. It is important to fill Mr. Frosty with 250 ml isopropanol and ensure that it is at RT before transferring to a −80°C freezer. After overnight cooling to −80°C, samples should be transferred to LN_2 without raising the sample temperature. When using Mr. Frosty or similar 1°C/min-freezing containers, extracellular ice is formed spontaneously at a random temperature, which introduces variability into the freezing outcome.

8 To maintain sterility, do not remove 0.5-cc straws from their bag. Just open one end of the bag where cotton-plugged ends of straws are located and label only

cotton-plugged ends of straws. Compared to cryovials, 0.5-cc straws are easy to seed extracellular ice and are preferable if small volumes (0.3–0.5 ml for each straw) of cell suspensions are cryopreserved. Nevertheless, similar viability rates were obtained after cryopreservation of ASCs in 0.5-cc straws and 1.5-cc cryovials (Lopez et al. 2016).

9 Use extreme caution when handling LN_2. Severe frostbite and even suffocation may occur as a result of direct LN_2 exposure and displacement of air by evaporating LN_2, respectively.

10 In our previous study, we compared thawing in air (RT) and in a water bath at 37°C after cryopreservation of ASCs in 0.5-cc straws. We did not observe any significant difference in terms of viability and plating efficiency (Lopez et al. 2016). However, we recommend thawing cryovials in a water bath at 37°C.

Reference

Lopez, M., R. J. Bollag, J. C. Yu, C. M. Isales, and A. Eroglu. 2016. "Chemically defined and xeno-free cryopreservation of human adipose-derived stem cells." *PLoS One* 11 (3):e0152161.

Cryopreservation of Red Blood Cells

Method I: High Glycerol/Slow Cooling Technique (Meryman and Hornblower 1972)

Preparation of the RBC Concentrate

Red blood cells (RBCs) can be prepared from whole blood using standard blood bank techniques. RBCs preserved in CPD (combined citrate/phosphate buffer and glucose containing whole blood/RBC anticoagulant/storage solution) or CPD-A1 (contains also adenine) may be stored at 1–6°C for up to 6 days before freezing. RBCs preserved in AS-1 and AS-S can be used as well. Both are glucose and adenine containing RBC-storage solutions, the latter (contains also a combined citrate/phosphate buffer) may be stored at 1–6°C for up to 42 days before freezing. Outdated RBCs that have undergone a rejuvenation procedure may be processed for freezing up to 3 days after their original expiration. RBCs in any preservative solution that have been entered for processing must be frozen within 24h after opening the system. The combined mass of the cells and the collection bag should be between 260 and 400g. Underweight units can be adjusted to approximately 300g either by adding isotonic saline or by removing less plasma than usual.

Addition of the Cryoprotective Solution

Warm the RBCs and the 6.2 M glycerol lactate solution to at least 25°C by placing them in a dry warming chamber for 10–15min or by allowing them to remain at room temperature for 1–2h. Otherwise the intracellular concentration of glycerol needed for successful protection of the RBCs might not be reached. Please note that RBCs of other species (e.g., bovine or canine) may have different temperature dependencies. The temperature must not exceed 42°C (at least in the case of human RBCs), otherwise hemolysis may take place. Place the container of RBCs on a shaker and add approximately 100ml of the

Preservation of Cells: A Practical Manual, First Edition. Allison Hubel.
© 2018 John Wiley & Sons, Inc. Published 2018 by John Wiley & Sons, Inc.

glycerol solution (6.2 M glycerol lactate solution as the red cells are gently agitated. For all transfer processes either a sterile tube docking device or plasma transfer sets should be used. Turn off the shaker and allow the cells to equilibrate, without agitation for 5–30 min. Allow the partially glycerolized cells to flow by gravity into the freezing bag. Cryogenic freezing containers may be made of, for example, polyolefin or polytetrafluorethylene. However, it is important that they are stabile under cryogenic temperatures, this includes especially the tubing when left at the bag. Add the remaining 300 ml of the glycerol solution in a stepwise fashion, with gentle mixing. Add smaller volumes of the glycerol solution for smaller volumes of red cells. The final glycerol concentration is 40% (w/v). Maintain the glycerolized cells at temperatures between 25 and 32°C until freezing. The recommended interval between removing the RBC unit from refrigeration and placing the glycerolized cells in the freezer should not exceed 12 h.

Cooling

Place the glycerolized unit in a cardboard or metal canister and place in a freezer at −65°C or below. The cooling rate should be less than 10°C/min. The temperature interval where this cooling rate should be measured has not been specified. Do not "bump" or handle the frozen cells roughly. Storage of the frozen RBCs at −65°C or colder is possible but not recommended for up to 10 years (and more).

Rewarming

For thawing, place the protective canister containing the frozen RBCs in either a 37°C water bath or a 37°C dry warmer. Agitate gently to speed up thawing. The thawing process takes at least 10 min, the thawed cells should be at 37°C.

Removal of the Cryoprotectant and Debris

After the RBCs are thawed, you may either use a commercial instrument (e.g., centrifuge) for batch or a continuous-flow washing device to deglycerolize cells. Follow the manufacturer's instructions precisely especially when using a special device (e.g., Cobe Cell Processor 2991). For batch washing, dilute the unit with a quantity of hypertonic (12%) sodium chloride solution appropriate for the size of the unit. Allow to equilibrate for approximately 5 min. Wash again with 1.6% sodium chloride until de-glycerolization is complete. Approximately 2 l of wash solution per unit are required. Suspend the de-glycerolized RBC I isotonic saline (0.9%) with 0.2% glucose. If you have opened the system for the processing, de-glycerolized RBCs must be stored at 1–6°C for no longer than 24 h (in the case of a transfusion).

Method II: A Low Glycerol/Rapid Cooling Technique (Rowe, Eyster, and Kellner 1968)

I) Preparation of the RBC concentrate: After collection of a unit of whole blood in ACD (citrate buffer and glucose containing whole blood/RBC/ platelet anticoagulant/storage solution, formulations ACD-A and ACD-B vary with regard to the concentration of the solutes) or CPD anticoagulant, the plasma is removed from the cells after centrifugation. The RBCs should be frozen as soon after the collection as possible, but preferably before it is 5 days old.

II) Addition of the cryoprotective solution: The remaining packed RBCs are weighed, and an equal volume by weight of the glycerol freezing solution is added at room temperature to achieve a final concentration of 14% (v/v). The freezing solution contains 28% (v/v) = 35% (w/v) glycerol, 3% mannitol, and 0.65% sodium chloride. After 14–30 min equilibration at room temperature (22°C), the RBC suspension is transferred into a suitable (e.g., polyolefin or polytetrafluorethylene) freezing bag. Conventional PVC plastic bags cannot be used, as they become brittle and crack upon freezing in liquid nitrogen (LN_2).

III) Cooling: The bag is placed between two metal plates (holder). These are used to keep the bag in a flat configuration. The top and bottom of the bag are tucked under to allow the holder plates to close without pinching the bag. Cooling is performed by complete immersion of the container in the open LN_2 filled without agitation. A freezer retainer should be used to prevent excess bulging of the container during freezing. Freezing is complete in 2–3 min when the LN_2 stops boiling. The unit is stored in its metal plate holder in a LN_2 storage tank either in the vapor or in the liquid phase.

IV) Rewarming: Upon retrieval from the LN_2 storage, the unit is immediately thawed by immersing the complete unit (bag and holders) into a 40–45°C warm water bath under gentle agitation (approximately 60 cycles/min) for about 2.5 min. The bag is removed from the metal holding plates and checked if all ice have disappeared. If not, immediately reimmerse the bag into the thaw bath, knead the bag under the warm water until the ice has completely melted.

V) Removal of cryoprotectant and debris: Following centrifugation and removal of the supernatant containing free hemoglobin, glycerol, and debris, the RBCs are gently washed three times using a bag centrifuge: The first wash is with 300–500 ml 3.5% sodium chloride at 4°C, the last two are with 1000–2000 ml isotonic saline (or preferably with 0.8% NaCl containing 200 mg/dl glucose). All washes must be added slowly to the cells at room temperature with gentle mixing. For resuspension of the RBCs after the de-glycerolization, the glucose supplemented sodium chloride solution may be used as well.

Method III: Hydroxyethylstarch/Rapid Cooling Technique (Sputtek 2007)

I) Preparation of the RBC concentrate: Approximately 450–500 ml of whole blood is collected in blood bank packs containing CPD-A as anticoagulant. Plasma and buffy-coat are removed using standard blood bank techniques and the RBCs are leuko-depleted by filtration and stored in an additive solution (SAG-M, adenine, glucose, and mannitol containing saline solution; PAGGS-S, phosphate buffer, adenine, glucose, guanosine containing saline solution; AS-1; AS-3; etc.) In the case of not leuko-depleted RBCs, five washing steps have to be performed to guarantee the same post-thaw saline stability as can be obtained with leuko-depleted RBCs after three washing steps. It is essential to remove white cells and platelets. White cells and platelets are highly thromboplastic, and as these cells are destroyed on freezing, thereby liberating their contents into the HES/RBC mixture, failure to filter could cause a disseminated intravascular coagulopathy when the RBCs are subsequently thawed and transfused without post-thaw washing. The leuco-depleted RBCs can be kept at $4 \pm 4°C$ for a maximum of 3 days prior to freezing. The first centrifugation takes place at $4000\,g$ for 10 min at 4°C. The supernatant additive solution is removed by means of a plasma extractor. The RBC concentrate is resuspended in 333 ml of an isotonic saline solution. The suspension is then centrifuged again. The centrifugation step is repeated three times to make sure that all of the additive solution and plasma have been removed and that the "contamination" of the RBC concentrates by leukocytes and platelets is minimal. After the removal of the last supernatant, purified RBC concentrates with a volume of approximately 220 ml at a hematocrit of $85 \pm 5.0\%$ should be obtained.

II) Addition of the cryoprotective solution: An equal part of the cryoprotective solution (CPS) containing 23% (w/w) HES and 60 mmol/l sodium chloride is added to the purified RBC concentrate whilst mixing continuous. The solution is commercially not available at present. It can be prepared from the dry HES powder or commercially available HES solutions after dialysis and freeze-drying by dissolving the dialyzed and freeze-dried HES and the appropriate amount of sodium chloride in distilled water. Suppliers of the dry HES powder are, for example, Ajinomoto (Japan) and Fresenius (Germany). Suppliers of HES solutions for infusion are, for example, Baxter (US), Fresenius (Germany), and Serum Werk Bernburg (Germany).

III) Two aliquots of 220 ml of this RBC suspension are transferred into two freezing bags. The bags must be carefully de-aerated and heat-sealed below the inlet port. As the densities of the CPS and the RBC concentrates are quite similar (ca. 1.08 g/ml), the final HES concentration in the suspension to be frozen is 11.5% (w/w). The two bags are then placed in two aluminum containers. The containers have a wall thickness of 2 mm. The exterior is pasted with a microporous textile tape to improve the heat transfer during the cooling process in boiling LN_2. Additionally, the closed containers produce a well-defined flat geometry of the bags and a homogenous sample thickness (approximately 5–6 mm). LN_2 is not allowed to come into contact with the samples during the initial cooling process. Please note that you will not be able to reproduce the results when not using the patented freezing container.

IV) Cooling: Cooling at the required cooling rate of about 240°C/min is achieved by complete vertical immersion of the containers into the open LN_2 filled Dewar. Use hand tongs, gloves, and goggles. Cooling is usually complete within 3 min. The total time is not critical as long as the containers are immersed for at least 3 min. Lift the reusable container from the LN_2. Open it quickly and remove the freezing bag. Transfer the freezing bag to the vapor phase of a LN_2 storage tank within 30 s to avoid the risk of premature thawing. The bags must be stored in vapor phase over the LN_2, not actually in the LN_2. Storage in the vapor phase over LN_2 below −130°C results in no time-dependent degradation.

V) Thawing: Rewarming is achieved by means of a shaking water bath. To guarantee thermally defined and reproducible conditions during rewarming, the bags have to be transferred from the LN_2 vapor phase into the pouch of the shaking water bath. This pouch maintains a well-defined flat geometry and effectively transduces the shaking frequency (300 cycles/min) and amplitude (2 cm) of shaking in the water bath (48°C) to the frozen unit. After 75 s, the temperature in the bag is about 20°C and they are immediately removed from the bath and the thawing container. Repeat this with the second bag prepared from this donation if needed.

VI) Removal of the cryoprotectant and debris: Both can easily be removed by washing the RBCs once with 300–500 ml of isotonic saline (or preferably with 0.8% NaCl containing 200 mg/dl glucose). Please note that HES should not be removed prior to transfusion (in contrast to glycerol) as long as a certain amount is not exceeded. This holds for "hemolysis" as well. However, when required, centrifuge refrigerated (4°C) at $4000 \times g$ for 10 min, remove supernatant by means of a plasma extractor. For resuspension of the washed RBCs, the glucose supplemented sodium chloride solution can be used. Standard RBC additive solutions (SAG-M, PAGGS-S, AS-1, and AS-3) may be used as well. The percentage "recovery" of intact cells, immediately post-thaw, is not used as a quality control procedure, because it is only a crude indicator of quality. Plasma stabilities may be slightly higher than saline stabilities because plasma factors allow a few very slightly damaged cells to recover. HES coats the surface of the RBCs and may provide a scaffolding for damaged membranes so that some cells appear intact, though they will rupture if diluted with isotonic saline. Viability in terms of "saline stability" can be determined as follows: 250 μl of the RBC suspension is diluted 40-fold in a buffered isotonic saline solution. After 30 min the suspension is separated into a supernatant (destroyed RBCs) and sediment (intact RBCs) by centrifugation. Saline stability is then calculated using the equation

$$\text{Saline stability}(\%) = (1 - \text{Hb}_S / \text{Hb}_T) \times 100,$$

where Hb_T corresponds to the total hemoglobin and Hb_S to the hemoglobin in the supernatant. The determination of the two hemoglobin concentrations can be performed spectrophotometrically at 546 nm using Drabkin's solution. A correction for the hematocrit is not required, as the volume fraction of erythrocytes after 40-fold dilution is less than 2%.

References

Meryman, H. T., and M. Hornblower. 1972. "A method for freezing and washing red blood cells using a high glycerol concentration." *Transfusion* 12 (3):145–156.

Rowe, A. W., E. Eyster, and A. Kellner. 1968. "Liquid nitrogen preservation of red blood cells for transfusion; a low glycerol-rapid freeze procedure." *Cryobiology* 5 (2):119–128.

Sputtek, A. 2007. "Cryopreservation of red blood cells and platelets." *Methods Mol Biol* 368:283–301.

Cryopreservation of Oocytes by Slow Freezing

Principle

Oocyte cryopreservation involves freezing unfertilized eggs from a patient for subsequent thawing, fertilization, and embryo transfer. Several types of patients may benefit from oocyte cryopreservation. Patients, especially single women, who are undergoing medical treatments, such as chemotherapy, may be at risk of irreversible loss of ovarian function. Such women may wish to freeze oocytes in an attempt to preserve fertility. Women undergoing Assisted Reproductive Technology (ART) procedures, but who have ethical or other concerns about embryo freezing, may opt to freeze oocytes.

Note: This procedure is no longer used to freeze oocytes since the change to vitrification of oocytes in 2007. It will remain in the procedure manual as a backup method and reference for oocytes frozen using this method.

Specimen Requirements

Oocytes are obtained after ovarian stimulation. These oocytes may be mature with first polar body extrusion, intermediate with no germinal vesicle, but no first polar body extrusion, or germinal vesicle stage immature oocytes. Oocytes can be frozen at any stage, but mature or GV stage oocytes are preferred. If GV stage oocytes are frozen, *in vitro* maturation of these eggs post-thawing will need to be accomplished before fertilization can be attempted.

Equipment and Supplies Needed

Equipment

(Freezing)
1) Planar embryo freezer including liquid nitrogen (LN$_2$) Dewar
2) Laminar flow hood

Preservation of Cells: A Practical Manual, First Edition. Allison Hubel.
© 2018 John Wiley & Sons, Inc. Published 2018 by John Wiley & Sons, Inc.

   3) Dissecting microscopes
   4) Analytical balance
   5) Liquid nitrogen storage tanks
   6) Pipette pump(Thawing)
   7) Storage tank containing oocytes to be thawed
   8) CO_2 incubator
   9) Stereoscopic microscope
   10) Inverted microscope with DIC optics

Supplies

(Freezing)
   1) Sodium-free oocyte freezing medium (SAGE Biopharma)
   2) 1.5 M PrOH in sodium-free media (SAGE)
   3) Falcon 3037 organ culture dishes
   4) Disposable Falcon 5 and 10 ml pipettes
   5) Pasteur pipettes
   6) Nunc cryovials (round bottomed, 1.8 ml capacity)
   7) Aluminum storage canes
   8) Cryosleeves
   9) Thermoprotective gloves
   10) Small plastic LN_2 Dewar
   11) 50 and 15 ml Falcon tubes
   12) Spatulas
   13) Marking pens(Thawing)
   14) Cryocanes containing vial(s) to be thawed
   15) Water bath warmed to 32°C
   16) Thermoprotective gloves
   17) 3037 Falcon organ culture dishes
   18) Thawing medium (SAGE Biopharma) 0.5 and 0.2 M sucrose solutions in sodium-free media
   19) 1 cc tuberculin syringes without needles
   20) Pasteur pipettes
   21) Culture drops for embryo culture (QFM + 10% SPS under oil)
   22) Stripper tips and device (150, 275 um)

Procedure

   A) Cryopreservation (Freezing)
      1) Before freezing, fill the Planar LN_2 Dewar with LN_2 leaving a little space at the top to allow evaporation. Place the pump into the Dewar and turn the Planar unit switch on. This provides power to the pump. Hit toggle switch on the bottom of the Dewar stand to pressurize the Dewar. The light will come on and go off automatically when Dewar is pressurized (above 5 psi on the dial on top of the Dewar).

2) Remove oocyte freezing media and the 1.5 M PrOH from the refrigerator and allow it to warm to room temperature.

3) Label cryovials with the date, patient name and ID, accession numbers, number of oocytes, and maturational stage of the oocytes. No more than three oocytes should be placed in any one vial, and oocytes of identical maturational stage should be placed in each vial (i.e., don't put a GV egg in the same vial as a mature egg).

4) Take 3037 dishes and mark one as PrOH and one as PrOH/SUCROSE with patient name.

5) Using a sterile Falcon pipette, pipette 1 ml of either PrOH or oocyte freezing media into separate 3037 dishes.

6) Using a sterile Falcon pipette, pipette 0.5 ml of oocyte-freezing media into each cryovial.

7) Using the Stripper device, transfer oocytes into the PrOH dish and hold at room temperature for 20 min. Watch the oocytes to be sure they shrink and re-expand. DO NOT leave the dish in the isolette during this step, as the cryoprotectant can be toxic at elevated temperature.

8) Using the Stripper device, transfer oocytes into the dish with oocyte freezing media containing 1.5 M PrOH/0.3 M sucrose. Rinse briefly in this dish and then pipette into vials.

9) Using the Stripper device, transfer oocytes into the cryovials. Rinse the pipette several times after each loading procedure to be sure that oocytes have been loaded into the vial.

10) Make sure the vials are tightly capped, then place the vials right side up in a vial holding rack and place it in the freezer. Make sure the vials stay right side up.

11) Turn the freezer on. Menu will come up and say SELECT MODE: RUN PROGRAM PRINT.

12) Hit RUN, and the screen will ask for a password. It is 3333. Enter the password.

13) Screen will show Run Program x, with the name of the program. Select the correct program by hitting the arrow keys up or down for the program number you want. Hit ENTER to select the correct program. For egg freezing, select Program 3: egg.

14) Hit RUN to start the program; this will bring the program to the correct starting temperature.

15) When the starting temperature has been reached, an alarm will sound and the screen will read "at start temp run." Hit CLEAR to silence the alarm. Place vials on a cane in the freezer chamber (no more than two vials per cane), and then hit RUN to begin the program.

16) The freezing program has several steps or ramps in the following order:
Ramp 1: cools at −2.0°C from the start temperature to −7.0°C.
Ramp 2: this is a hold step and is shown on the display as a cooling rate of +0.0°C; the vials are seeded during this time.

Ramp 3: cools at −0.3°C/min to −35°C.

Ramp 4: a hold for 20 min at −35°C.

17) At approximately 15–20 min after starting the program, the program will switch from ramp 1 to Ramp 2, the seeding ramp. Ramp 2 is a hold ramp, where the samples are held at −6.0°C for 15 min. You want to seed after the vials have been held at this temperature for 5 min. An alarm will sound when it is time to seed. Hit CLEAR to silence the alarm, and then seed each vial.

18) For seeding, wear cryogloves and a face shield to fill the plastic LN_2 Dewar with LN_2, and place the large forceps into the Dewar to cool. When ready to seed, put on a thermoprotective glove and grab the cooled forceps. Bring the cane holding the vial to be seeded out of the freezing chamber, and touch the cooled forceps to the outer wall of the vial at the level of the meniscus. You should see the outer wall of the vial become cloudy as ice begins to form at the point where the vial was touched. When you see this, immediately return the cane to the inside of the freezing chamber.

19) Follow the above procedure for all the vials to be seeded.

20) After all vials have been seeded, hit Run to start the program again.

21) Toward the end of Ramp 2, check to see that all the vials have been seeded by removing them briefly from the chamber—the solution within the vial should be forming ice. Note: this should be done quickly to prevent the samples from warming above the seeding temperature.

22) Make sure the program goes into Ramp 3, which will cool the samples at −0.3 to −35°C.

23) Once the program has entered Ramp 4 (a hold at −35°C), remove the canes from the freezer and IMMEDIATELY plunge the vials into the small LN_2 Dewar containing LN_2. The entire vial should be plunged and covered with LN_2.

24) Using thermoprotective gloves and a face shield, transfer vials to storage canes, which are labeled on top with the date and patient name. You can put a maximum of three vials per cane.

25) Place the cane into a cardboard sleeve. The cardboard sleeve should be labeled with the date of freeze, patient name, accession numbers of the oocytes stored on the cane, and the cell stage of the stored oocytes.

26) Transfer the cardboard sleeve to a storage tank.

27) Allow the chamber to warm to room temperature before turning the freezer off. The LED display will read "Ready to Restart" when it is OK to turn the freezer off.

28) Take a blank oocyte cryopreservation sheet from the cryo notebook in the lab and fill in all the required information for each oocyte. Make a copy of the sheet and keep the copy in the lab notebook. The original is kept in the patient chart. Also, fill in information in the accession log (yellow spiral log book) kept in the lab. The date, patient name, number and stage of oocytes frozen, accession numbers, and canister/tank stored are recorded here.

29) Make sure you fill in the cryo notebook in the IVF lab for each freeze. A sample report is attached, and the same format should be followed for each entry.

B) Thawing

1) The day before thawing, pull the copy of the cryo sheet from the logbook in the IVF lab and place on the chart, and pull out the corresponding page in the cryo logbook. This is done to be sure you will thaw the correct oocytes.

2) Timing of the thaw:
   a) Natural cycles-patients will be given an hCG injection to time the thaw in these cycles; thaw oocytes on the day of hCG.
   b) Estrace/progesterone replacement cycles: Oocytes are thawed on the first day of progesterone.

3) Remove thaw solutions from refrigerator and allow it to warm to room temperature before use.

4) Adjust water bath temperature to 32°C.

5) Remove cane from the storage tank using thermoprotective gloves and remove vial(s) to be thawed.

6) Quickly unscrew the cap of the vial to vent any LN_2 gas in the vial—this is important to do so that the vial does not explode when thawing.

7) Tighten the cap and immerse the vial in the 32°C water bath, making sure that the entire contents of the vial are immersed.

8) Allow the vial to thaw for approximately 1–1.5 min, or until you see that there is no ice left in the vial.

9) Transfer the vial to the hood, and using a Pasteur pipette pipette the contents of the vial into the inner well of a 3037 dish.

10) Examine the contents of the dish for oocytes.

11) If all oocytes are not recovered, flush the vial with 0.5 M sucrose and repeat step 10.

12) Move the embryos into a 3037 dish containing 1 ml of 0.5 M sucrose in sodium-free media.

13) Hold at room temperature for 10 min.

14) Move oocytes to dish with 1 ml of 0.2 M sucrose in sodium-free media and hold at room temperature for 10 min.

15) Wash oocytes in the outer well of a culture dish prepared on the previous day (and equilibrated overnight) then transfer to the center drop and place in the incubator for culture.

16) About 30 min after thawing, remove oocytes from the incubator and examine under the inverted microscope to determine if oocytes have survived thaw. Degenerated/nonviable oocytes will have lost the integrity of the plasma membrane and will appear dark and granular.

17) If there are one or less oocytes that have survived the thaw, then thaw more vial(s) until there are four to six intact oocytes if possible. Decisions on how many oocytes to thaw should be made with the patient and physician before starting the thaw.
18) Call the patient to inform them of the number thawed and survived, and notify the physician office.
19) Culture oocytes for a minimum of 3 h post-thaw, then inseminate with intracytoplasmic sperm injection (ICSI). Inseminate using ICSI regardless of sperm quality and follow standard IVF procedures.
20) Fill out the cryo form, indicating the date of thaw, person performing the thaw, number thawed and survived, and outcome. Make sure you indicate the number of oocytes that remain frozen.

Safety

1) Liquid nitrogen will burn skin upon contact. Make sure thermoprotective gloves and a face shield are worn when working with LN_2.
2) The LN_2 in the Dewar attached to the embryo freezer is under pressure. DO NOT REMOVE the pump from the LN_2 Dewar unless the pressure is at zero. If you remove the pump while the Dewar is pressurized, LN_2 may explode and burn the operator. Also, exposure to a large amount of LN_2 can cause asphyxiation.
3) If LN_2 contacts the skin, immediately place the affected part under water to warm the area. In the event of a large leak or exposure to a large surface area, medical attention should be sought immediately.

Calculations

None

Reporting Results

1) Results are to be filled out on the cryo form, in the cryo notebook, and in the cryo log book.
2) The physician office should be contacted and the patient contacted to let them know the outcomes of the freeze/thaw. For patients, this can be done either by a phone call or by informing the patient directly at the time of embryo transfer.

Procedure Notes

1) Selection of oocytes for freezing is critical for optimal performance. Selection criteria are listed under SPECIMEN REQUIREMENTS earlier.
2) The Planar freezer is on emergency backup power, thus the likelihood of an electrical failure is minimal. If the freezer breaks down or does not operate correctly during a run, use the following steps:
    a) If a problem is noted before seeding, then remove the vials, warm to room temperature, then step the embryos out of cryoprotectant (i.e., do a thaw procedure) and place in the incubator for continued culture until the freezer is repaired.
    b) If the problem occurs after seeding, then plunge the vials into LN$_2$ and store as per usual.
    c) Contact the patient and physician office to notify them as to what has occurred.

Limitations of Procedure

Oocyte freezing is an experimental procedure and no guarantees or data can be given as to success rates with thawing, fertilization, or chances of pregnancy.

Oocyte Vitrification and Warming

Principle

Vitrification is an alternative method for cryopreservation of oocytes and embryos. In vitrification, cells are placed into relatively high concentrations of cryoprotectant, such that upon cooling no intracellular ice is formed. Instead, the interior of the cell forms a glass-like substance. The theoretical advantage of vitrification is that the formation of intracellular ice is avoided, which if not controlled properly can lead to formation of large intracellular ice crystals that cause irreversible damage to the cell. For vitrification to be effective, several conditions must be met. One, cells must be exposed to a very high concentration of cryoprotectant. Second, exposure time of cells to cryoprotectant must be limited, since prolonged exposure to the high concentrations of cryoprotectant used in vitrification can become toxic and cause cell death. Third, after treatment with cryoprotectant cells must be plunged directly into liquid nitrogen (LN_2). Fourth, warming must take place very rapidly in order to avoid damage to the cells. If vitrification is done correctly then survival rates post-thaw of greater than 90% have been consistently reported in the literature.

Equipment and Supplies

Equipment

1) Liquid nitrogen storage tanks
2) Analytical balance
3) Nikon dissecting microscope, non-heated stage

Supplies

1) Modified HTF (SAGE)
2) Serum protein substitute (SPS) (SAGE)

Preservation of Cells: A Practical Manual, First Edition. Allison Hubel.
© 2018 John Wiley & Sons, Inc. Published 2018 by John Wiley & Sons, Inc.

3) Vitrification kit (SAGE) containing
   - Equilibration solution (7.5% DMSO, 7.5% ethylene glycol)
   - Vitrification solution (15% DMSO, 15% ethylene glycol)
4) Cryoleaf (Origio)
5) Cryocane
6) Plastic goblet fitted into cryocane
7) Cryosleeves
8) Stripper tips and device (150 μm, 275 μm)
9) 3003 dish lids (retrieval dishes)
10) Liquid nitrogen Dewar
11) Small Styrofoam container for plunging of leafs/straws
12) 0.2 μm syringe filters
13) 10–30 ml syringes
14) Falcon 15 ml conical tubes
15) 5–10 ml pipettes
16) Variable volume pipettes with tips (10–100 μl)
17) Sucrose (Sigma)
18) Weighing paper
19) Weighing spatula

Procedure

A) Vitrification (Freezing) of Oocytes
   1) *Details for oocytes*
      a) Only mature oocytes with a defined first polar body should be vitrified.
      b) Select oocytes with no cytoplasmic defects-oocytes with vacuoles or bull's eye granular appearance should not be vitrified.
      c) Oocytes should be vitrified within 1 h and no more than 2 h after egg retrieval to avoid prolonged culture time. Currently available data suggests that embryo quality is adversely affected if oocytes are frozen for more than 2 h after retrieval.
   2) *Preparation of solutions*
      The vitrification kit from SAGE contains the two premade solutions needed for vitrification. The solutions should be allowed to reach room temperature prior to use. Alternatively, the solutions can be prepared in the following manner:
      a) Vitrification equilibration solution (7.5% DMSO/7.5% EG)
         i) Make up stock solution of mHTF with 20% SPS by combining 8 ml mHTF + 2 ml SPS
         ii) Add 8.5 ml of mHTF/20% SPS to a Falcon 15 ml tube
         iii) Add 0.75 ml DMSO to the tube and mix well

    iv) Add 0.75 ml EG to the tube and mix well

    v) Filter through a 0.2 μm filter and store refrigerated until use— expiration date is 6 months from preparation

  b) Vitrification solution (15% DMSO, 15% EG, 0.5 M sucrose)

    i) Weigh out 1.71 g sucrose and add to 15 ml Falcon conical tube

    ii) Add 7 ml mHTF/20% SPS and dissolve sucrose

    iii) Add 1.5 ml DMSO and mix thoroughly

    iv) Add 1.5 ml EG and mix thoroughly

    v) Filter sterilize through a 0.2 μm filter and store refrigerated until use, expiration date is 6 months from preparation

3) Method

Note: All steps are done at room temperature. Use the stereoscope on the bench top for all procedures.

  a) Label all cryoleafs/straws with the date, patient name, date of birth, number of oocytes stored, and the accession numbers. No more than two oocytes should be frozen per cryoleaf/straw. Label a cryocane, goblet, and outer plastic cryosleeve with the patient information described above.

  b) Complete an Oocyte Cryopreservation Log with the necessary information and write the patient's information in the accession number log book.

  c) Prepare drops of media as shown in Figure 1.

    Drop 1: 50 μl mHTF with 20% SPS

    Drop 2: 75 μl mHTF (20% SPS) + 25 μl of SAGE equilibration solution (ES), yielding a 25% ES drop

    Drop 3: 50 μl mHTF (20%SPS) + 50 μl ES, yielding a 50% ES drop

    Drop 4: 100 μl of ES solution (100% ES)

    Drops 5–7: 100 μl drops of vitrification solution (VS)

  d) Rinse no more than four oocytes at a time in drop 1.

  e) Pipette oocytes into drop 2 and hold for **3 min**. They should shrink and then recover almost all of their original volume.

  f) Pipette oocytes into drop 3 and hold for **3 min**. Watch for shrinkage and then re-expansion as they recover their volume.

Figure 1 Vitrification freezing diagram.

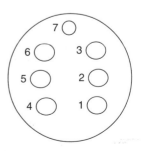

g) Pipette oocytes into drop 4 and hold for **6 min**. Watch to ensure they shrink initially and then recover their volume. They should recover 80–90% of their volume by the end of the 6-min period and should appear as a normal, rounded shape. During this time you should fill the Dewar as well as the Styrofoam holding container with LN$_2$.

h) After 6 min in drop 4, observe the oocytes. If they do not appear to have recovered their volume to 80–90% of original and/or do not appear rounded, then hold for an additional minute and reexamine. When the oocytes appear as described above, then move to the next steps. You can hold in drop 4 for up to 9 min. Proceed with further steps at 9 min regardless of appearance, but make a note on the freezing sheet if oocytes have not recovered by the 9-min mark.

i) Using a 175-μm Stripper tip, transfer a **maximum of two** oocytes to drop 5 of vitrification solution. NOTE: The eggs will immediately float due to the higher density of the vitrification solution.

j) Pipette up and down several times in drop 5 of vitrification solution to ensure equilibration The oocyte(s) will undergo dramatic shrinkage during this time.

k) Using the same Stripper tip move oocytes into drop 6 of vitrification solution then use a second 175 μm Stripper tip and move into drop 7 of vitrification solution.

l) After **40 s from initially placing into drop 5** (first drop of vitrification solution), pick up oocytes and keep at the very end of the Stripper tip to minimize volume of media used during vitrification.

m) Pipette eggs onto the cryoleaf while observing under stereomicroscope. Try to keep volume to a minimum to the point where the oocytes are somewhat flattened in appearance-remove any excess media with Stripper tip if necessary.

n) Immediately plunge the cryoleaf into the Styrofoam container filled with LN$_2$, and then slide the protective green cover over the tip of the leaf. While keeping the vitrified sample under LN$_2$, insert the leaf containing the oocytes into the labeled outer sheath.
Note: Place outer sheath in the LN$_2$ before doing this, so it is filled with LN$_2$ when sample is transferred.

o) QUICKLY transfer vitrified leafs/straws into a goblet in the LN$_2$ Dewar. Cover with the pre-labeled cryosleeve and transfer to storage tank.

p) Fill in the log book in the lab, as well as a cryopreservation form for each patient. Keep one copy of the form in the lab notebook and one in the patient's chart.

B) Warming (Thawing)

1) *Preparation of solutions*

a) Use vitrification warming kit from SAGE. The kit has a vial with 1.0 M sucrose, a vial with 0.5 M sucrose, and a vial with isotonic warming solution. Modified media supplemented with 20% SPS will be used in place of the included isotonic (MOPS) warming solution. Alternatively, 1.0 and 0.5 M sucrose can be prepared as outlined in the following.

b) Prepare mHTF w/20% SPS: In a 15 ml conical tube add 2 ml SPS to 8 ml mHTF. Label the tube and reserve for 0.5 M sucrose preparation and remainder for thaw process.

c) Prepare 1 M sucrose: Add 3.4 g sucrose to a second tube of 10 ml mHTF/20% SPS and label appropriately.

d) Prepare 0.5 M sucrose: In a 15 ml conical tube add 5 ml of prepared 1 M sucrose (above) and 5 ml of prepared mHTF with 20% SPS.

e) Filter sterilize the sucrose solutions by passage through a 0.2 μm filter and store refrigerated until use—expiration date is 6 months from preparation.

2) Method

a) Before beginning, remove the aforementioned reagents from the refrigerator and bring to room temperature. Prepare dishes for thaw as outlined in the following (Figure 2).

All steps are to be performed at 37°C.

- Dish 1: Add 1 ml of 1 M sucrose to the inner well of a 3037 dish and place in the isolette so it comes to a temperature of 37°C.
- Dish 2: In a 3002 dish prepare four drops as follows:
  Drop 1: 50 μl of 0.5 M sucrose for wash
  Drop 2: 50 μl of 0.5 M sucrose to hold for 2 min
  Drop 3: 50 μl of mHTF w/20% SPS for wash
  Drop 4: 50 μl of mHTF w/20% SPS to hold for 3 min
  Cover with warmed, equilibrated mineral oil and place in isolette.

b) Working quickly, move canes from the storage tank to a Dewar filled with LN$_2$. Fill a Styrofoam container with LN$_2$ to hold the cryoleafs during the thaw.

c) Quickly transfer the leafs/straws to be thawed into the Styrofoam container. Move the Styrofoam container as close to the warmed isolette as possible.

Figure 2 Vitrification thawing diagram.

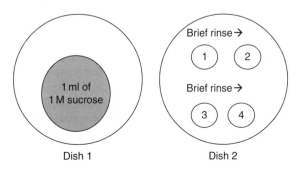

Dish 1   Dish 2

d) While **completely under LN₂**, remove the outer sheath from the leaf to be thawed and slide back the protective green cover exposing the tip containing the oocytes.

e) Remove leaf from the LN₂ and IMMEDIATELY place it into the warmed 1 M sucrose solution in the 3037 dish while looking through the stereomicroscope. *Do not delay in plunging the leaf or straw because the sample will not warm appropriately and the oocyte will be damaged.*

f) You should see the oocytes float off the leaf/straw into the solution. If it appears the oocytes are stuck to the holder, it may be necessary to pipette them off with a stripper tip.

g) Using the end of the cryoleaf, gently swirl the 1.0 M sucrose to completely expose the oocytes to the 1.0 M sucrose.

h) Keep the oocytes in the warmed 1 M sucrose for **1 min**. Observe the oocytes for shrinkage of the cytoplasm to about 50% of the normal volume. The plasma membrane should have a very sharp appearance, appearing as a distinct black line when observed under the stereoscope. If you do not see this then hold the oocytes as necessary in the warm 1.0 M sucrose for a **maximum of 2 min**.

i) Aspirate some media from Drop 1, then pick up the oocytes from the 1.0 M sucrose and transfer them to Drop 1. Aspirate media from Drop 2 and transfer the oocytes from Drop 1 to Drop 2. Hold for **2 min**.

j) Aspirate media from Drop 3, then move the oocytes from Drop 2 to Drop 3 for a brief wash. Transfer from Drop 3 to Drop 4 and hold for **3 min**.

k) Rinse in fertilization media with **20% SPS** and culture in incubator for a minimum of 3 h before ICSI.
   Note: The fertilization media for warming of vitrified oocytes should be supplemented with *20% SPS* for culture post-thaw for a maximum of 4 h, then moved to media supplemented with 10% for the remaining culture period.

3) Complete all paperwork for IVF chart on number thawed and survived, and update cryopreservation log books and computer database.

Quality Control

1) Only use reagents and products that have passed either the Mouse Embryo or Human Sperm Motility QC tests, either in-house or by the manufacturer.

2) Only use media that is within the expiration date given by the manufacturer.

Safety

Be extremely careful when handling LN_2. Cryogloves and face shields are available in the lab and should be used when working with LN_2 to avoid burns. In case of a large spill, evacuate the area as large amounts of LN_2 can cause asphyxiation.

Transportation of Hematopoietic Progenitor Cells and Other Cellular Products

Principle/Rationale

Hematopoietic progenitor cell (HPC) and other cellular products must be transported in such a manner as to ensure cellular viability and function. Many of these components are essential for a patient's survival, as this type of patient undergoes high-dose marrow ablative treatment prior to transplant. If a patient has received myeloablative therapy, it is essential that only knowledgeable, trained individuals transport components from the collection center to the transplant center. These components must be packaged both to protect their integrity during transport and the health and safety of individuals in the immediate area.

Specimen

1) HPC products
2) Therapeutic cells
3) Other products

Equipment/Reagents

1) Validated thermally insulted shipping container (cooler) labeled as biohazardous
2) Plastic zip lock biohazard bags
3) Absorbent material such as Chux.
4) Frozen gel packs
5) Room temperature gel packs
6) Refrigerated gel packs
7) Temperature recording unit less than $-130°C$ or equivalent
8) Monitoring device for -30 to $85°C$

Preservation of Cells: A Practical Manual, First Edition. Allison Hubel.
© 2018 John Wiley & Sons, Inc. Published 2018 by John Wiley & Sons, Inc.

9) Liquid nitrogen (LN$_2$) dry shipper
10) Appropriate insulated container for shipping via Fed Ex either at 4°C or at room temperature (20–24°C)
11) Alcohol wipes

Quality Control

1) LN$_2$ dry shipper containing a temperature monitoring device.
2) The LN$_2$ dry shippers are validated twice per year to ensure that they are holding temperature less than –140°C for at least 72 h.

Procedure

A) **For Transportation of Products within Contiguous Facilities**
  1) Disinfect the inside and outside of a validated transport container.
  2) Place product in a plastic zip lock biohazard bag and seal to prevent leakage.
  3) Package product in cooler:
     a) For thawed products:
        1) Disinfect two refrigerated gel packs with alcohol wipes.
        2) Place product on a refrigerated gel pack.
        3) Wrap product and refrigerated gel pack in absorbent material and place in cooler.
        4) Place a second refrigerated gel pack on top of product.
     b) For fresh products:
        1) Disinfect two room temperature gel packs with alcohol wipes.
        2) Place on a room temperature gel pack.
        3) Wrap product and room temperature gel pack in absorbent material and place in the cooler.
        4) Place a second room temperature gel pack on top of the product.
B) **Transportation of Cryopreserved Products**
  1) Prior to shipping cryopreserved product charge a dry shipper. Charging a dry shipper takes a minimum of 48 h.
  2) Coordinate with receiving facility how product will be shipped (courier service, FedEx, etc.) and how the empty dry shipper will be returned.
  3) Visually inspect dry shipper for integrity and ensure that it could withstand leakage of contents, shocks, pressure changes, and other conditions incident to ordinary handling in transportation.
  4) Ensure that temperature is below –150°C.
  5) Verify product label against the patient identifiers. Place into charged dry shipper.

a) Place cassettes or vials in a plastic bag.
b) Do not remove the outer metal cassettes from products that are being shipped.
c) Place absorbent and packing material in dry shipper around product so that product is snug.
d) Ensure that product does not remain at room temperature for longer than 2 min.
6) Place product in the center of dry shipper and close the inner lid.
7) Ensure that all records required for transport are included.
a) Close the lid and secure using zip ties, tape, or other means.
8) Upon return of dry shipper:
a) Reinspect integrity and ensure that it could withstand leakage of contents, shocks, pressure changes, and other conditions incident to ordinary handling in transportation.
b) Download temperature history of the container from the temperature recording unit (when applicable).
C) **Transit Time**
1) For all products, the transit time must be adequate to ensure product safety.
2) For all products, the transit time should ensure adequate time will remain upon arrival for processing and/or infusion.

Additional Information

A) If a recipient is myeloablated, HPC's shall be hand-carried by a courier who has received instruction in transportation requirements.
B) HPC products **MUST NOT BE IRRADIATED** and therefore must not be passed through any type of x-ray device during transport.
C) If commercial transportation is used such as airline transportation, fresh products must be carried in a temperature and pressure-controlled compartment. These products shall be hand-carried, packaged as indicated in the aforementioned procedure by a person who is qualified by training to transport this type of product.
D) Fresh products must be transported in a rigid, puncture-proof container and NEVER left unattended.

Further Reading

AABB Standards for Cellular Product Services, 8th ed., 2017.
FACT-JACIE International Standards for Cellular Therapy Product Collection, Processing, and Administration, 6th ed., 2015.

Cryopreservation of Hematopoietic Progenitor Cells

Principle/Rationale

This procedure describes the cryopreservation of hematopoietic progenitor cell (HPC), Apheresis [HPC(A)] products using a closed system via volume reduction, utilizing manual centrifugation and plasma expression. Proper technique during the processing and cryopreservation procedures will ensure adequate viability and engraftment potential. Cells will be cryopreserved at a concentration not exceeding 5.0×10^8 total cells/ml prior to the addition of cryoprotectant solution. After the addition of cryoprotectant solution, HPC(A) products will be aliquoted into approved freezing bags for cryopreservation in a controlled-rate freezer (CRF), then transferred into the vapor phase of liquid nitrogen (LN_2) tanks for long-term storage.

Preservation of Cells: A Practical Manual, First Edition. Allison Hubel.
© 2018 John Wiley & Sons, Inc. Published 2018 by John Wiley & Sons, Inc.

Protocol/Processing Schema

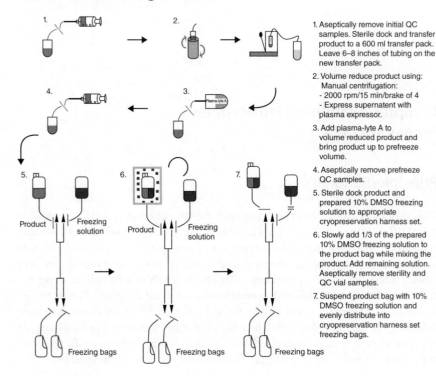

1. Aseptically remove initial QC samples. Sterile dock and transfer product to a 600 ml transfer pack. Leave 6–8 inches of tubing on the new transfer pack.

2. Volume reduce product using:
   Manual centrifugation:
   - 2000 rpm/15 min/brake of 4
   - Express supernatent with plasma expressor.

3. Add plasma-lyte A to volume reduced product and bring product up to prefreeze volume.

4. Aseptically remove prefreeze QC samples.

5. Sterile dock product and prepared 10% DMSO freezing solution to appropriate cryopreservation harness set.

6. Slowly add 1/3 of the prepared 10% DMSO freezing solution to the product bag while mixing the product. Add remaining solution. Aseptically remove sterility and QC vial samples.

7. Suspend product bag with 10% DMSO freezing solution and evenly distribute into cryopreservation harness set freezing bags.

Specimen

1) HPC(A)
2) HPCs marrow

Equipment/Reagents

1) 3 ml sterile syringes
2) 10 ml sterile syringes
3) 20 ml sterile syringes
4) 60 ml sterile syringes
5) 18 gauge needle
6) EDTA purple top vacutainer tubes
7) Cryostore freezing bag set or equivalent
8) 300 ml transfer pack

9) 600 ml transfer pack
10) 2 ml Nunc vials
11) Antimicrobial Susceptibility Testing (AST) and nocistatin, (NST) culture bottles
12) Sampling site coupler
13) Dispensing Pin with Luer Lock Valve
14) Clave connector
15) Mini spike
16) Dimethylsulfoxide (DMSO)
17) 25% Human serum albumin (HSA)
18) 500 ml Plasma-lyte A
19) Alcohol wipes
20) Iodine pads
21) Metal freezing cassettes
22) Hand heat sealer
23) Calibrated scale
24) Freezing blankets
25) Controlled-rate freezer (CRF)
26) J6-MI centrifuge with JS 4.2 rotor
27) Terumo Steri-welder with wafers
28) Plasma expressor
29) Biological safety cabinet (BSC)
30) Hemostat
31) Biohazard bags

Quality Control

1) All reagents must pass visual inspection for discoloration and/or turbidity, and supplies should be checked for integrity.
2) Sterility testing is performed at the receipt of product and at the end of processing.
3) Total nucleated cell (TNC) counts are performed initially and prior to adding freezing solution. TNC recovery is calculated.
4) Sampling for CD34 counts is taken prior to adding freezing solution.

Procedure

A) Set-up and Labeling of Paperwork, Freezing Bags Cassettes, and cryovials
   1) Label each freezing cassette with the patient and product identifiers.
   2) Place appropriately labeled reagents and supplies in the BSC as needed for processing. All equipment used in processing should be cleaned before use.

3) Transfer a bag of Plasma-lyte A into a 600 ml Transfer Pack (TP) labeled with an in-process label, and write Plasma-lyte A on this TP-600 with initial and date.

B) Receipt of Product
1) Place an appropriate tare bag on the scale, tare the scale, and then weigh the product on the tarred scale.

C) Collection of QC samples
1) Place the product into a cleaned BSC. Spike the bag with a dispensing pin and clave connector.
2) Gently mix the product well.
3) Aseptically remove samples for QC testing.

D) Centrifugation of Product
1) Aseptically sterile dock and then transfer the HPC (A) product to a 600 ml transfer pack labeled with an in-process label. Once completed, use a hemostat to clamp the 600 ml transfer pack containing the product.
   Note: If excess product remains in the original collection bag after transfer, aseptically add Plasma-lyte A to the original collection bag to rinse the bag, and then transfer the remainder of the cells into the transfer pack.
2) Aseptically sterile dock the 600 ml transfer pack containing Plasma-lyte A to the 600 ml transfer pack containing the product and ensure the hemostat is secure.
3) Tare a scale using a TP-600 tare bag and place the product bag on the scale.
4) Transfer Plasma-lyte into the product bag to bring the volume up to within 10% of 500 ml for centrifugation. Leave 6–8 inches of tubing tail on the product bag, and heat seal off the Plasma-lyte.
5) Use an alcohol wipe to clean the Plasma-lyte bag and return to BSC for later use.
6) Place the transfer pack containing the product into a sterile biohazard bag and place in a centrifuge spin cup. Weigh the product and counter balance with aTP-600 filled with 500 ml of Plasma-lyte A.
7) Centrifuge the product for 15 min at 2000 RPM (824 RCF) with a brake of 4 (6 min from 500 RPM to zero) at room temperature.
8) After centrifugation, carefully remove the product from the centrifuge and place on the plasma expressor. Be careful not to disturb the pellet.
9) Sterile dock the product tail to a new TP-600 transfer pack. Express the plasma. Stop the flow when the cells approach the shoulder of the bag. Also, take note of any wisps of cells that start to break off from the main pellet. *Take care not to lose any cells.*
10) Heat seal three times and disconnect the tubing, using the middle seal, making sure to leave a 6–8 inch tail on the concentrated HPC(A) product.

Note: Do not discard waste bag until an acceptable recovery has been determined. Additional Total Nucleated Counts, (TNC) counts may be required if the Quality Indicator is not met.

11) Gently resuspend the concentrated product by massaging the corners and mixing well to minimize clumping.

12) Tare a scale using a TP-600 tare bag and place the product bag on the scale to weigh.

13) Disinfect the product and return to BSC. Spike the product bag with a dispensing pin and clave connector.

E) Pre-freeze preparation

1) Using aseptic technique, bring up the product volume to half of the total volume to be frozen by injecting Plasma-lyte A with a sterile syringe.

   a) For all products, ensure the maximum concentration of cells does not exceed the limit of the bag being used for cryopreservation.

2) Resuspend the concentrated product with the appropriate amount of Plasma-lyte A and **mix well**. Examine for clumping.

   Note: If platelet clumping is present, allow product to sit at room temperature for 5 min and then resuspend a second time by mixing.

3) Gently mix the product and aseptically remove samples for QC testing using a dispensing pin and clave connector.

F) Preparation of DMSO Freezing Solution for a Final Concentration of 10%

To determine total volume of reagents required, multiply the desired numbers of bags by the values listed in Table 1.

1) Insert a sampling site coupler into the 300 ml transfer pack. (Note: Do not use a dispensing pin with Luer Lock Valve and clave connector.)

2) Using a sterile syringe and needle, add the appropriate volume of Plasma-lyte to the labeled 300 ml transfer pack.

3) Draw up the appropriate volume of DMSO and inject quickly while mixing into the transfer pack containing Plasma-lyte.

4) Refrigerate the DMSO Plasma-lyte solution at 4°C for at least 10 min before attaching to cryopreservation harness set and adding 25% HSA.

5) The appropriate volume of 25% HSA *will be added to the solution immediately prior to cryopreservation*. See Table 1 for HSA volumes.

Table 1 Preparation of DMSO freezing solution for a final concentration of 10% vol/vol.

Reagent	Per 50 ml	Per 100 ml
Volume of DMSO	5	10
Volume of plasma-lyte A	10	20
Volume of 25% HSA	10	20

G) Cryopreservation
1) Turn on the CRF.
2) Remove the DMSO freezing solution from refrigerator.
3) Sterile dock the pre-freeze product bag and prepared DMSO freezing solution to cryopreservation harness set (clamp lines leading to freezing bags). Wrap both in a freezing blanket and place in BSC. DO NOT REMOVE THE PRODUCT FROM THE FREEZINGBLANKET.
4) Place labeled freezing cassettes into a freezing blanket and place in BSC.
5) Add 25% HAS, refer to Table 1 for volumes of HSA to be added to the DMSO freezing solution (10%) via a sterile syringe and needle.
6) Unclamp the lines, slightly raise the freezing solution bag and slowly add contents (via gravity) to the pre-freeze product while gently mixing the chilled product bag.
7) Once all the freezing solution has been added to the pre-freeze bag, clamp the line to the product bag containing the added freezing solution.
8) Mix well, aseptically remove samples for QC.
9) Hang the product bag and open all clamps to the freezing bags and check that there are no kinks in the tubing. Allow the product to flow down and evenly distribute into freezing bags.
   a) If three or more bags are to be frozen, remove appropriate volume (50 or 100 ml) from the product bag via 60 ml syringe and inject into each freezing bag from a double harness set via the port.
   Note: If freezing bag harness set is not available, refer to schematic in Appendix A.
10) Inspect for bag and product integrity (i.e., no clots or leaks) and ensure even distribution of product. Heat seal off tubing to freezing bags three times and separate using the middle seal.
   Note: To minimize bag breaks, leave a small tail while sealing. Leave enough space to create a "safety seal."
11) Place the products in the labeled freezing cassettes. Transfer product bags and vials for cryopreservation using a validated transport container.
12) Place the freezing cassettes and Nunc vials into the CRF and initiate the appropriate freezing program.
13) Once the CRF has completed, immediately transfer the freezing cassettes and vials to the LN_2 storage tanks (vapor phase).
14) Review the CRF chart to assure instrument ran appropriately. Look for dips or peaks in the chamber or probe curve that could indicate that the run was less than ideal.

Appendix A   Alternate Cryopreservation Harness Set-2 or 4 Bags

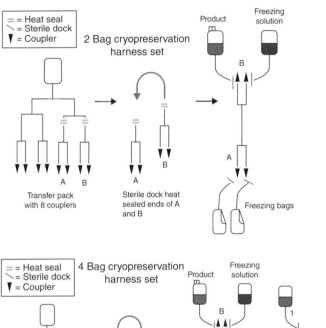

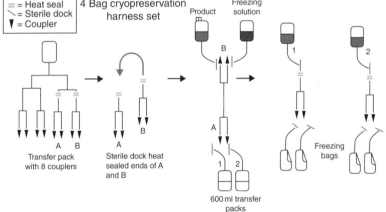

Further Reading

AABB Technical Manual, 18th ed., 2014.

AABB Standards for Cellular Therapy Product Services, 8th ed., 2017.

FACT-JACIE International Standards for Cellular Therapy Product Collection, Processing, and Administration, 8th ed., 2015.

Validating a Closed System for Hematopoietic Progenitor Cell Cryopreservation Utilizing Centrifugation and Sterile Docking Techniques. Study Number 8488754.

Thawing of Hematopoietic Progenitor Cells

Principle/Rationale

The purpose of this procedure is to define the correct protocol for thawing for infusion hematpoietic progenitor cells from apheresis [HPC(A)] and bone marrow [HPC(M)] products which have been processed, cryopreserved in bags, and stored in vapor phase of liquid nitrogen (–150°C or less).

Equipment/Reagents

1) Sterile bags
2) Water bath
3) Distilled water
4) 300 ml transfer pack
5) Sebra heat sealer
6) Baker Biological Safety Cabinet (BSC)
7) ACD-A
8) 3, 10, or 20 ml sterile syringe
9) 18 gauge sterile needle
10) Sampling site coupler
11) Povidone iodine preps
12) Alcohol prep pads
13) 3 ml EDTA tube
14) AST and NST culture bottles
15) Touch plates
16) Hemostat

Preservation of Cells: A Practical Manual, First Edition. Allison Hubel.
© 2018 John Wiley & Sons, Inc. Published 2018 by John Wiley & Sons, Inc.

Quality Control

1) All reagents must pass visual inspection for discoloration and/or turbidity, and supplies should be checked for integrity.
2) Sterility and viability testing is performed at the end of processing.

Procedure

A) Prepare water bath and BSC for thaw procedure
   1) Clean water bath and prepare for daily use.
   2) Clean BSC.
   3) Place appropriately labeled reagents and supplies in the BSC as needed for processing. All equipment used in processing should be cleaned before use.
   4) Place a volume of ACD-A into the transfer pack that is equal to 10% of the volume being thawed:
      a) Determine the amount of ACD-A which is to be added by checking the order and processing record. The volume of ACD-A, which will be added, is 10% of the volume being thawed.
      b) Using a hemostat, clamp off the tubing on the TP-300.
      c) Aseptically insert a sampling site coupler into one of the ports.
      d) Disinfect the port of the ACD-A and the sampling site coupler with alcohol.
   5) Using appropriately sized syringe and needle, transfer the appropriate volume of ACD-A to the TP-300.
B) Thaw product
   1) Remove product(s) from storage using a validated container. Once in the lab, verify the correct product has been retrieved.
   2) A primary technologist will thaw the product.
      a) Note time and water bath temperature.
      b) The expiration time is one and a half hours (90 min) from the time that the cryopreserved product enters the water bath and is recorded both on the label and on the photocopy of the label as soon as possible.
      c) Place the freezing cassette with product into a thaw bag and submerge in the water bath until frost is not apparent on the cassette.
      d) Remove the unit from the cassette and verify parent number on bag is correct product to be thawed. Place the unit in a new thaw bag and complete the thaw in the water bath. Gently mix product until thawed.

3) Once product is completely thawed, remove it from the water bath and transport to the BSC.
   a) Transfer the product from the cryobag to the TP-300 by completing the following:
      i) Wipe the port of the cryobag with an alcohol wipe.
      ii) Use the spike on the tubing of the TP-300 to spike the port on the cryobag.
      iii) Transfer product from cryobag into TP-300.
      iv) Place hemostat on tubing close to the TP-300.
      v) If thawing more than one bag then
         a) Wipe the port of the additional cryo bag with alcohol wipe.
         b) Carefully remove the spike on the TP-300 line from the thawed bag.
         c) Spike the additional bag with and transfer product into the TP-300.
         d) Wipe the sampling site coupler with alcohol wipe.
   b) Aseptically remove samples for QC testing.

Further Reading

AABB Standards for Cellular Product Services, 8th ed., 2017.
AABB Technical Manual, 19th ed., 2017.
FACT-JACIE International Standards for Cellular Therapy Product Collection, Processing, and Administration, 6th ed., 2015.

Processing and Cryopreservation of T-Cells

Principle/Rationale

The goal of processing the mononuclear cells (MNCs) obtained by apheresis is to enumerate CD3$^+$ T-cells and dose them for infusion or cryopreservation. After enumeration of CD3$^+$ T-cells by flow cytometry the MNC, Apheresis product is known as T-cells, Apheresis.

Protocol/Processing Schema: N/A

Specimens

1) MNC, apheresis [MNC(A)]
2) T-cells, apheresis

Equipment/Reagents

1) 3 ml sterile syringes
2) 10 ml sterile syringes
3) 20 ml sterile syringes
4) 60 ml sterile syringes
5) 18 gauge needle
6) EDTA purple top vacutainer tubes
7) Cryostore freezing bag set or equivalent
8) 300 ml transfer pack
9) 600 ml transfer pack
10) 2 ml Nunc vials
11) AST and NST culture bottles
12) Sampling site coupler
13) Dispensing Pin w/Luer Lock Valve

Preservation of Cells: A Practical Manual, First Edition. Allison Hubel.
© 2018 John Wiley & Sons, Inc. Published 2018 by John Wiley & Sons, Inc.

14) Clave connector
15) Mini spike
16) Dimethylsulfoxide (DMSO)
17) 25% human serum albumin (HSA)
18) 500 ml plasma-lyte A
19) Alcohol wipes
20) Iodine pads
21) Metal freezing cassettes
22) Hand heat sealer
23) Calibrated scale
24) Freezing blankets
25) Controlled-rate freezer (CRF)
26) J6-MI centrifuge with JS 4.2 rotor
27) Terumo Steri-welder with wafers
28) Plasma expressor
29) Biological safety cabinet (BSC)
30) Hemostat
31) Biohazard bag

Quality Control

1) All reagents must pass visual inspection for discoloration and/or turbidity, and supplies should be checked for integrity.
2) Sterility testing is performed at the receipt of product and at the end of processing.
3) TNC counts are performed.
4) Sampling for CD3.

Procedure

A) Set-up and Labeling
   1) Label each freezing cassette with the patient and product identifiers.
   2) Place appropriately labeled reagents and supplies in the BSC as needed for processing. All equipment used in processing should be cleaned before use.
   3) Using a ring stand and clamp, set up an inverted 60 ml syringe and a three-way stopcock.
B) Receipt of Product
   Place an appropriate tare bag on the scale, tare the scale, and then weigh the product on the tarred scale.

C) Collection of QC samples
   1) Place the product into a cleaned BSC. Spike the bag with a dispensing pin and clave connector.
   2) Gently mix the product well.
   3) Aseptically remove samples for QC testing.
D) Cryopreservation and Storage
   1) Media preparation
      a) Preparing DMSO solution (freezing solution)
         i) Transfer 60 ml Plasma-lyte A into a sterile 250 ml conical tube. With a 10 ml pipette and pipette aid, add 15.0 ml of DMSO into Plasma-lyte (volumes can be doubled if needed).
         ii) Store the solution at 1–6°C until ready to use.
      b) Preparing Plasma-lyte/HSA solution (freezing media)
         i) Transfer 60 ml Plasma-lyte A into a properly labeled sterile 250 ml conical tube and add 30 ml of 25% HSA (volumes can be doubled if needed).
         ii) Store at 1–6°C until ready to use.
   2) Cryopreservation
      a) Dose the bags for cryopreservation and separate the doses into 250 ml conicals.
      b) Centrifuge the product for 15 min at 2000 RPM (824 RCF) with a brake of 9 (30 s from 500 RPM to zero) for 15 min.
         i) Take care when handling conical post-centrifugation so as not to disturb the pelleted cells.
         ii) Disinfect and return to BSC.
      c) Uncap and aspirate supernatant using a sterile 25 ml pipette or 10 ml pipette and pipette aid without disturbing cell pellet.
      d) Resuspend cell pellet by pipetting up and down with a 10 ml pipette and pipette aid until cell solution appears homogenous.
      e) Add the appropriate amount of Plasma-lyte/HSA solution to the resuspended cell pellet in the 250 ml conical (refer to Table 1).
      f) Place each 250 conical(s) in a metal cup(s) surrounded by frozen plastic ice cubes or equivalent.

Table 1 Proper volume of cryopreservation and plasma-lyte/HSA solutions for different bag capacities.

# Bags	Bag type (ml)	Plasma-lyte/HSA solution (ml)	Plasma-lyte/DMSO solution (ml)	Total volume (ml)
1–4	50	18.75	18.75	37.5
1–4	100	37.5	37.5	75.0

g) Using a pipette aid with a 10 ml pipette drip the appropriate volume of Plasma-lyte/DMSO (refer to Table 1) into the 250 ml conical while gently swirling the product.
h) Remove QC samples.
i) Attach the cryobag to the previously prepared ring stand and inverted syringe.
Note: Ensure that the stopcock valve is in the closed position.
j) Pour the product into the freezing bags via the inverted 60 ml syringe.
k) Once product is in bag, close line with a hemostat and heat seal the tubing with three seals. Repeat this procedure for additional bags as needed.
l) Aseptically remove QC samples.
3) Inspect for bag and product integrity (i.e., no clots or leaks) and ensure even distribution of product. Heat seal off tubing to freezing bags three times and separate using the middle seal.
Note: In an effort to minimize bag breaks, leave a small tail while sealing. Leave enough space to create a "safety seal" in the event that you experience a sealing failure.
4) Place the products in the labeled freezing cassettes. Transfer product bags and vials for cryopreservation using a validated transport container.
5) Place the freezing cassettes and Nunc vials into the controlled-rate freezer (CRF) and initiate the appropriate freezing program.
6) Once the CRF has completed, immediately transfer the freezing cassettes and vials to the liquid nitrogen storage tanks (vapor phase).
7) Review the CRF chart to assure instrument ran appropriately. Look for dips or peaks in the chamber or probe curve that could indicate that the run was less than ideal.

Further Reading

AABB Standards for Hematopoietic Progenitor Cell and Cellular Product Services, 8th ed., 2017.
Areman, E., J. Deeg, and R. Sacher. 1992. Bone Marrow and Stem Cell Processing, 292–323, Philadelphia, PA: E.A. Davis.
FACT-JACIE International Standards for Cellular Therapy Product Collection, Processing, and Administration, 6th ed., 2015.
JC, Comprehensive Accreditation Manual for Laboratory and Point of Care, 1st ed., 2017.
Ritz, J. and S.E. Sallan. July 10, 1982. "Autologous BMT in CALLA positive ALL after in vitro treatment with J5 monoclonal antibody and complement," *Lancet* 2:60–63.

Thawing and Reinfusion of Cryopreserved T-Cells

Principle/Rationale

Cryopreserved T-cells, apheresis must be thawed and washed prior to infusion to best optimize recovery, viability, and remove DMSO from the product.

Protocol/Processing Schema

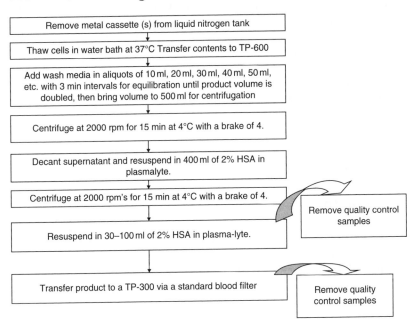

Preservation of Cells: A Practical Manual, First Edition. Allison Hubel.
© 2018 John Wiley & Sons, Inc. Published 2018 by John Wiley & Sons, Inc.

Specimen

Cryopreserved T-cells, apheresis

Equipment/Reagents

1) Biological safety cabinet (BSC)
2) Heat sealer
3) 60 ml syringe
4) 20 ml syringe
5) 3 ml syringe
6) 18 gauge safety needles
7) 21 gauge safety needles
8) Sterile thawing bags
9) Plasma transfer set
10) Plasma-lyte A
11) 25% HSA
12) 50 ml ACD-A
13) Preservative free heparin, 1000 units/ml for IV
14) 70% isopropyl alcohol swabs
15) Iodine
16) 600 ml transfer pack
17) 300 ml transfer pack
18) J6 MI centrifuge
19) Water bath
20) Dispensing pin
21) Clave connector
22) NST and AST sterility bottles
23) Plasma expresser

Quality Control

1) All reagents must pass visual inspection for discoloration and/or turbidity, and supplies should be checked for integrity.
2) Sterility, TNC, and viability testing is performed at the end of processing.

Procedure

A) **Preparation of Materials**
   1) Place appropriately labeled reagents and supplies in the BSC as needed for processing. All equipment used in processing should be cleaned before use.

2) Prepare water bath.
3) Set centrifuge to 4°C so that it has time to cool prior to centrifugation of product.

B) **Preparation of Thawing Media**
1) Make the thawing media (Table 1) by performing the following:
   a) Label a TP-600 with "Plasma-lyte/ACD/25% HSA/Heparin."
   b) Transfer 360 ml of Plasma-lyte to a TP-600.
   c) Add 45 ml of ACD-A into the TP-600.
   d) Add 67.5 ml of 25% HSA into the TP-600.
   e) Add 12 ml of Heparin into the TP-600
   f) Refrigerate the media at 1–6°C until ready for use.
2) Make a 2% HSA in Plasma-lyte solution (Table 1):
   a) Label a TP-600 with "2% HSA in Plasma-lyte"
   b) Transfer 460 ml of Plasma-lyte to the TP-600.
   c) Add 40 ml of 25% HSA to the TP-600.
   d) Refrigerate the media at 1–6°C until ready for use.

C) **Thawing Cryopreserved Product**
1) Transport product to lab using a validated container.
2) Once in lab verify that correct product has been retrieved.
3) Place cryotin in sterile thaw bag.
4) Place cryotin in 37°C water bath until frost is gone from the outside of the tin.
5) Place product in BSC and remove product from thaw bag and cryotin.
6) Place cryopreserved product into a sterile thawing bag.
7) Place thawing bag with cryopreserved product into 37°C water bath and record the time and temperature. Gently rock back and forth until completely thawed.
8) Dry the thawing bag and disinfect prior to placement in BSC. Remove the product from thawing bag.
9) Transfer product into a 600 ml transfer pack (TP-600) by performing the following:
   a) Disinfect a refrigerated gel pack, place it in the BSC and place labeled TP-600 on it.
   b) Disinfect port on product bag and spike with the line from the TP-600, transfer thawed product.

Table 1 Preparation of thawing media.

	Plasma-lyte A (ml)	ACD (ml)	25% HSA (ml)	Heparin at 1000/ml
Thawing media	360	45	67.5	12
2% HSA in plasma-lyte	460	—	40	—

10) Place a hemostat on the line close to the product.
11) Heat seal three times and disconnect the tubing, using the middle seal, making sure to leave a 6–8 inch tail.

D) **Wash Thawed Product**
   1) Sterile dock the "Thawing Media" prepared in Section B to the TP-600 containing the product.
   2) Double the product volume by adding thawing media to the product:
      a) Place product on tarred scale to measure volume of thawing media as it is added.
      b) Remove hemostat between two bags and slowly add 10 ml of thawing media to the product and clamp the line between the two bags using a hemostat.
      c) Place product on a refrigerated gel pack and gently rock for 3 min.
      d) Repeat steps **D.2.a** through **D.2.c** adding 20, 30, 40, and 50 ml of thawing media until the original volume of the product has been doubled.
   3) Tare a scale with an empty TP-600.
   4) Place product on scale and remove hemostat between bags.
   5) Add thawing media until the product has reached a volume of 500 ml.
   6) Separate the bags using the heat sealer and leave an adequate amount of tubing on the TP-600 containing the product.
   7) Centrifuge the product for 15 min at 2000 RPM (824 RCF) with a brake of 4 (6 min from 500 RPM to zero) at 4°C.
   8) Express supernatant
      a) Carefully remove product from centrifuge and secondary container in a manner that it does not disturb the pellet.
      b) Hang on plasma expresser.
      c) Place a hemostat on the tubing of the TP-600 containing the product.
      d) Sterile dock the product tail to a new TP-600 transfer pack, labeled with "Waste, Not for Infusion."
      e) Express the plasma.
      f) Stop the flow when the cells approach the shoulder of the bag.
      g) Separate the bags using a heat sealer leaving 8–10 inches of tubing.
   9) Resuspend the cells
      a) Resuspend the concentrated product and examine for clumping.
      b) Weigh the concentrated product on an appropriately tarred scale.
      c) Sterile dock the bag of 2% HSA/Plasma-lyte solution onto the product bag.
      d) Tare a scale with an empty TP-600.
      e) Place product bag on scale and slowly bring the product volume to 400 ml with 2% HSA/Plasma-lyte mixture.
      f) Separate the bags using a heat sealer, leaving 8–10 inches of tubing on the product bag.

10) Centrifuge the product
  a) Place product in a secondary container, such as a small zip lock biohazard bag.
  b) Prepare a balance that is within 10 g of the weight of the product.
  c) Centrifuge the product for 15 min at 2000 RPM (824 RCF) with a brake of 4 (6 min from 500 RPM to zero) at 4°C.
11) Express supernatant
  a) Carefully remove bag from centrifuge and secondary container in a manner that does not disturb the pellet.
  b) Gently hang product on plasma expresser.
  c) Place a hemostat on the tubing of the TP-600 containing the product.
  d) Sterile dock the product tail to a new TP-600 transfer pack, labeled "Waste, Not for Infusion."
  e) Slowly express the plasma into the waste bag.
  f) Stop the flow when the cells approach the shoulder of the bag.
  g) Separate the bags using the heat sealer, leaving an adequate tail.
12) Resuspend the cells
  a) Resuspend the concentrated product.
  b) Mix well and examine for clumping.
  c) Tare a scale with an empty TP-600.
  d) Weigh the concentrated product on the scale.
E) **Preparation for Product Infusion**
  1) Dilute the concentrated cells pellet with a small amount of 2% HSA/Plasma-lyte solution.
  2) Transfer the product to a TP-300 via a standard blood filter.
  3) To rinse the bag, aseptically add a minimum of 30 ml of additional 2% HSA/Plasma-lyte solution to the post spin bag and then transfer the contents through the filter.
  4) Add additional 2% HSA/Plasma-lyte to the concentrated product to bring it to a final total volume of 52 or 102 ml.
  5) Aseptically remove samples for QC testing.

Further Reading

AABB Standards for Cellular Product Services, 8th ed., 2017.

FACT-JACIE International Standards for Cellular Therapy Product Collection, Processing, and Administration, 6th ed., 2015.

Food and Drug Administration, Human Cells, Tissues, and Cellular and Tissue-Based Products; Donor Screening and Testing and Related Labeling (21 CFR Part 1271), 2005.

JC, Comprehensive Accreditation Manual for Laboratory and Point of Care, 1st ed., 2017.

Index

Preservation of Cells: A Practical Manual, First Edition. Allison Hubel.
© 2018 John Wiley & Sons, Inc. Published 2018 by John Wiley & Sons, Inc.